CAMBRIDGE LIBRARY COLLECTION

Books of enduring scholarly value

Life Sciences

Until the nineteenth century, the various subjects now known as the life sciences were regarded either as arcane studies which had little impact on ordinary daily life, or as a genteel hobby for the leisured classes. The increasing academic rigour and systematisation brought to the study of botany, zoology and other disciplines, and their adoption in university curricula, are reflected in the books reissued in this series.

Paxton's Flower Garden

Best remembered today for his technically innovative design for the Crystal Palace of 1851, Joseph Paxton (1803–65) was head gardener to the Duke of Devonshire at Chatsworth by the age of twenty-three, and remained involved in gardening throughout his life. Tapping in to the burgeoning interest in gardening amongst the Victorians, in 1841 he founded the periodical *The Gardener's Chronicle* with the botanist John Lindley (1799–1865), with whom he had worked on a Government report on Kew Gardens. *Paxton's Flower Garden* appeared between 1850 and 1853, following a series of plant-collecting expeditions. Only three of the planned ten volumes were published, but with hand-coloured plates (which can be viewed online alongside this reissue) and over 500 woodcuts, the work is lavish. Further colour plates of orchids are to be found in Volume 2, clearly a reflection of Lindley's interest, but also of the wider fascination for these flowers.

Cambridge University Press has long been a pioneer in the reissuing of out-of-print titles from its own backlist, producing digital reprints of books that are still sought after by scholars and students but could not be reprinted economically using traditional technology. The Cambridge Library Collection extends this activity to a wider range of books which are still of importance to researchers and professionals, either for the source material they contain, or as landmarks in the history of their academic discipline.

Drawing from the world-renowned collections in the Cambridge University Library, and guided by the advice of experts in each subject area, Cambridge University Press is using state-of-the-art scanning machines in its own Printing House to capture the content of each book selected for inclusion. The files are processed to give a consistently clear, crisp image, and the books finished to the high quality standard for which the Press is recognised around the world. The latest print-on-demand technology ensures that the books will remain available indefinitely, and that orders for single or multiple copies can quickly be supplied.

The Cambridge Library Collection will bring back to life books of enduring scholarly value (including out-of-copyright works originally issued by other publishers) across a wide range of disciplines in the humanities and social sciences and in science and technology.

Paxton's Flower Garden

Volume 2

Joseph Paxton
John Lindley

CAMBRIDGE
UNIVERSITY PRESS

CAMBRIDGE UNIVERSITY PRESS

Cambridge, New York, Melbourne, Madrid, Cape Town,
Singapore, São Paolo, Delhi, Tokyo, Mexico City

Published in the United States of America by Cambridge University Press, New York

www.cambridge.org
Information on this title: www.cambridge.org/9781108037266

© in this compilation Cambridge University Press 2011

This edition first published 1851-2
This digitally printed version 2011

ISBN 978-1-108-03726-6 Paperback

PAXTON'S

FLOWER GARDEN.

BY

PROFESSOR LINDLEY AND SIR JOSEPH PAXTON.

VOL. II.

LONDON :
BRADBURY AND EVANS, 11, BOUVERIE STREET.
1851-2.

PLATE 37.

L. Constans, del. & Lith.

Printed by C.F.Cheffins, London.

[PLATE 37.]

I. THE BROWN BISTORT.

(POLYGONUM BRUNONIS.)

II. THE BILBERRY-LEAVED BISTORT.

(POLYGONUM VACCINIIFOLIUM.)

A Pair of handsome Hardy Herbaceous Plants, from NEPAL, *belonging to the Natural Order of* BUCKWHEATS.

Specific Characters.

I. THE *BROWN* BISTORT.—Stems perennial, creeping, rising up at the points. Leaves smooth, oblong-lanceolate, narrowed to the base, slightly serrated. Spikes terminal, narrow, somewhat broken, solitary.

POLYGONUM *BRUNONIS;* caulibus perennibus reptantibus ascendentibus, foliis glabris oblongo-lanceolatis basi angustatis serrulatis, spicis terminalibus elongatis interruptis solitariis.

Polygonum Brunonis : *Wallich Cat., No.* 1692. *Meisner in Wallich Plantæ Asiaticæ rariores,* III., 54. *Royle, Illustrations of the Himalayas, t.* 80, *f.* 3.

II. THE *BILBERRY-LEAVED* BISTORT. — Stems perennial, creeping, ascending. Leaves ovate, tapered to each end, shining, quite entire. Spikes long, thin, with the lower flowers distant from each other.

POLYGONUM *VACCINIIFOLIUM ;* caulibus perennibus reptantibus ascendentibus, foliis ovatis utrinque attenuatis lucidis integerrimis, spicis elongatis tenuibus floribus inferioribus remotis.

Polygonum vacciniifolium : *Wallich Catalogue, No.* 1695. *Meisner in Plant. As.,* III., 54. *Royle's Illustrations, t.* 80, *f.* 2.

THESE two beautiful little plants belong to the same race as our own charming wild Bistort, and represent it on the mountains of N. India. Dr. Royle says they occur on such mountains as Mussooree, Choor, Kedarkanta, and Peer Punjal, or from 7000 to near 13,000 feet of elevation, and on the Kherang, Chandow, and other passes, along with the European viviparous Bistort (*P. viviparum*). This seems to be the greatest elevation at which such plants occur in India.

As might be expected, they are perfectly hardy in this country, living without protection in low cold damp soils, provided they are placed among stones, so that no wet may lodge about them in winter. In such situations they rapidly cover the ground with a bright verdure formed of their entangled prostrate stems; and in the early autumn throw up an abundance of their gay rosy or rich brown red spikes. Hence they are particularly well adapted to rockwork, when very hardy species are required.

Of the two the Bilberry-leaved is the more delicate, and requires to be kept clear of all neighbouring plants that may choke it up. Its favourite place is with its back to a lump of rock, and a gentle slope before it to the south or west. Not trailing so much as the other, it forms a little flat loose bush, through which the air and light pass partially.

But the Brown Bistort lies flat upon the earth, and with vigorous arms easily thrusts aside any invaders of its territory. We see it now (Feb. 22) piercing the winter mat formed by its old dead leaves, and struggling upwards into renewed growth.

Both are propagated readily by cuttings put into light damp earth in the month of August beneath a handglass.

PLATE 38

L.Constans.Pinx.& Zinc

Printed by C.F.Chefflins London

[Plate 38.]

THE ANDERSON VERONICA.

(VERONICA ANDERSONII.)

———•———

A GARDEN HYBRID.

♂ Veronica speciosa.   ♀ Veronica salicifolia.

════════════════

EVERY one is now familiar with the graceful half-hardy Willow-leaved Veronica of New Zealand (*V. salicifolia*), which is so commonly known in gardens under the aliàs of *V. Lindleyi*. It is a shrub with narrow willow-like leaves, and graceful spikes of white flowers.

Equally common has become that far more imposing, though less graceful species, the Showy Veronica or Napuka (*V. speciosa*), from the same country, conspicuous for its broad blunt solid leaves imitating those of the laurel, and its stiff erect massive tufts of deep violet flowers.

Both are nearly hardy—in Devonshire and Cornwall quite so: but in most of the English and all the Scotch counties they can only be made to thrive as greenhouse plants, among which they are the gayest in their season, which is autumn.

It occurred to a very intelligent gentleman named Anderson, residing at Maryfield, near Edinburgh, who has long occupied himself with questions of hybridisation, that the two plants would probably cross. He therefore powdered the stigma of the Willow-leaved Veronica with the pollen of the Napuka, from which sprang a most extraordinary race, now called *V. Andersonii*. The leaves of the mule are broader than those of the Willow-leaved, narrower and thinner than those of Napuka; the manner of growth and form of the tufts of flowers is exactly intermediate, and

B 2

stranger still each tuft is particoloured, white at the bottom like the Willow-leaved, rich violet at the top like the Napuka! In short, the newly constituted plant is one of the most beautiful of all those which the art of man has yet, with all reverence be it said, succeeded in creating. It flowers copiously in September and October.

We understand that Mr. Anderson has yet another curiosity of this kind, between *Veronica saxatilis* and *V. fruticulosa*, blue, with purple veins " through a distinct intermediate."

PLATE 39

L. Constans del &Lith

Printed by C.F.Cheffins, London.

[PLATE 39.]

I. THE SPOTTED PLEIONE.

(PLEIONE MACULATA.)

II. THE BOTTLE PLEIONE.

(PLEIONE LAGENARIA.)

Terrestrial Alpine Herbaceous Plants, *from* NORTHERN INDIA, *belonging to the Natural Order of* ORCHIDS.

Specific Characters.

I. THE *SPOTTED* PLEIONE. — Pseudobulbs whole-coloured, short, thick, rounded, narrowed at the base. Bract short, inflated, roundish, hooded. Sepals and petals oval, acute. Lip rounded, entire, emarginate, with 7 crested lines.

I. PLEIONE *MACULATA ;* pseudobulbis unicoloribus teretibus crassis basi angustatis, bracteâ brevi subrotundâ cucullatâ inflatâ, sepalis petalisque ovalibus acutis, labello rotundato integerrimo emarginato lineis 7 cristatis.

Gomphostylis candida : *Wallich ic. ined.* Cœlogyne maculata : *Lindl. in Wallich, Plant. As. rariores*, I., 45, *t.* 53.

II. THE *BOTTLE* PLEIONE.—Pseudobulbs flask-shaped, clouded. Bract hooded, acute, very much tapering to the base. Sepals and petals linear-lanceolate, acuminate. Lip rounded, entire, emarginate, with 5 crested lines.

II. PLEIONE *LAGENARIA ;* pseudobulbis lagenæformibus nebulosis, bracteâ cucullatâ acutâ basi longe angustatâ, sepalis petalisque lineari-lanceolatis acuminatis, labello rotundato integerrimo emarginato lineis 5 cristatis.

THESE beautiful plants are from the Alps of India. The first was found by Dr. Wallich's collectors on rocks and the trunks of trees, among moss, on the Khasija Mountains. Mr. Thomas Lobb found both in such places, and sent them to Messrs. Veitch, to whom we are indebted for specimens received last October. The Khasija name, according to Dr. Wallich, is *Atia-Chakarpate.*

The Spotted Pleione has long been known to botanists as a species belonging to that Alpine group of so-called Cœlogynes, of which *C. Wallichiana* is the best known example, and to which the *Epidendrum præcox* and *humile* of Smith are also assigned. They certainly resemble greatly the genus *Cœlogyne*, to which one of us, many years since, reduced them, but they differ in certain points, to which we shall advert when we proceed to figure *Pleione humilis,* another charming species, of which Messrs. Veitch have also obtained possession.

The Bottle Pleione was received at the same time, mixed with the last. It is, however, obviously distinct, not only in colour, but in the form of the bracts and lobes of the flower, and in the number of crested lines upon the lip. It was distributed among Dr. Wallich's dried plants, No. 1763, under the name of *Cœlogyne humilis.*

Those who have seen the wondrous beauty of the Wallich Pleione at Chatsworth, will easily understand what these may become under similar treatment. The method followed in cultivating such mountaineers in Bengal is thus stated by Dr. Wallich:—

" They were introduced into the Hon. East India Company's Garden at Calcutta, in 1816, and I have often had the satisfaction of seeing them in flower there. Our mode of treating these and similar *Epiphytes* is to place them on beds made of brickwork, raised four or five feet from the ground, containing a rich mould, mixed with a large proportion of pebbles, and resting on a stratum of large stones or masses of vitrified bricks, so as to admit of being perfectly drained. The surface is covered with a quantity of moss, and the whole structure is placed in a shady and sheltered situation, corresponding to the natural place of the growth of such plants. By the aid of these beds, and by a constant attention to the necessity of keeping the roots as well as the plants themselves moderately moist, I have succeeded in cases even where there was but little hope ; for instance, with plants from the higher regions of Nepal, and even from Gossain Than, in the Himalaya."

All the species are Alpine. Dr. Royle describes the *P. præcox* as being found ornamenting with its large, richly-coloured flowers the branches of Oaks on Loudour, at 7500 feet of elevation, in 30° N. Lat., but only during the moisture of the rainy season. The secret of their successful cultivation in England lies in keeping them cool and dry while at rest, and forcing them with heat, moisture, and bright light as long as they are inclined to grow. What plants are they for exhibition tables !

GLEANINGS AND ORIGINAL MEMORANDA.

249. DEUTZIA GRACILIS. *Zuccarini*. A fine hardy deciduous shrub from Japan, with weeping branches profusely covered with small white flowers. Belongs to the Syringas (Philadelphaceæ). Introduced by Dr. v. Siebold. (Fig. 121.)

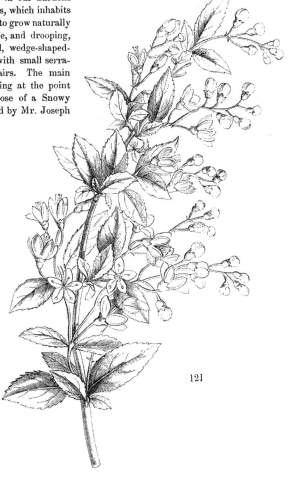

The beauty of such Deutzias as we have already in our Gardens renders every new species an object of interest. This, which inhabits the damp valleys and lofty mountains of Japan, is said to grow naturally about two yards high; its branches are long, flexible, and drooping, especially when in flower. The leaves are small, wedge-shaped-lanceolate, or ovate-lanceolate, tapering to the point with small serratures, and a coating on both faces of fine starry hairs. The main branches are covered with lateral branchlets, bearing at the point graceful racemes of white flowers about as large as those of a Snowy Medlar (*Amelanchier*). The stock has been purchased by Mr. Joseph Baumann, of Ghent, who will put it into circulation. It is said to prefer a soil that is slightly moist in summer.

It is necessary to observe that under the false name of Deutzia gracilis, there exists in Gardens a worthless Callicarp, which Siebold calls *C. Murasaki.*

250. CAMPANULA COLORATA. *Wallich.* (*aliàs* Campanula Moorcroftiana, *Wallich.*) A trailing frame perennial? with purple flowers of little beauty. Native of the Sikkim-Himalaya, and other mountainous Indian regions. Introduced at Kew.

Raised from seeds sent by Dr. Hooker, to the Royal Gardens of Kew, in 1849, from Sikkim-Himalaya, at an elevation of 10,000 feet above the level of the sea. It seems quite hardy, and flowered through the autumn in the open border, even as late as November. It is variable in its growth, sometimes erect, sometimes trailing. " Its copious deep-coloured bell-flowers would render it a great ornament for rock-work." From its appearance at the present time, we have every reason to think it will prove perennial. Few plants from the elevated regions within or near the tropics are able to resist the severity of some of our winters without

121

protection. It is therefore desirable to keep plants in small pots, under a frame, planting them out in the spring.—*Bot. Mag.*, t. 4555.

On the opposite page are figures of some little known species of Oncidium, viz:—

251. ONCIDIUM LUNATUM. *Lindley, in Bot. Reg.*, t. 1929. Flowers pale primrose, with rich brown spots; lip white, with pale brown stains.—Demerara.—(Fig. 122, *about natural size.*)

252. ONCIDIUM GRACILE. *Lindley, in Bot. Reg.*, 1920. Flowers whole-coloured, yellow.—Brazil.—(Fig. 123, *twice the natural size.*)

253. ONCIDIUM SPHEGIFERUM. *Lindley, in Bot. Reg.*, 1843, *misc.* 23. Flowers very pale clear greenish yellow, with the sepals and petals stained with rust at the base; lip clear yellow, with numerous broken crimson bands.—Brazil.—(Fig. 124, *flower, twice the natural size.*)

254. ONCIDIUM SERPENS. *Lindley, Genera et Sp. Orch.*, p. 204. Flowers yellow, spotted with dark brown.—Peru.—(Fig. 125, *flower, about natural size.*)

255. ONCIDIUM PULVINATUM. *Lindley, in Bot. Reg.*, 1838, *misc.* 115. Flowers bright yellow, with a crimson base to the sepals and petals, and numerous specks of the same colour on the lip. —Brazil.—(Fig. 126, *flower, less than natural size.*)

256. ONCIDIUM WENTWORTHIANUM. *Bateman, in Bot. Reg.*, 1840, *misc.* 194. Flowers yellow, with deep brown bars on the sepals and petals, and a cinnamon-coloured stain over the base of the lip.—Guatemala.—Of this there are two distinct varieties of size and colour; the second, in the possession of Sir Philip Egerton, has flowers twice as large and as richly coloured as in the variety first known. (Fig. 127, *flower, natural size of the original variety.*)

257. ONCIDIUM DELTOIDEUM. *Lindley, in Bot. Reg.*, t. 2006. Flowers bright yellow, whole-coloured, except the lip and column-wings, which are spotted with rich red.—Peru.—(Fig. 128, *flower, natural size.*)

258. ONCIDIUM SUTTONI. *Bateman, in Bot. Reg.*, 1842, *misc.* 8. Flowers greenish-yellow, with the base of all the parts a uniform dirty brown.—Mexico.—(Fig. 129, *a flower, rather above the usual size.*)

259. ONCIDIUM NANUM. *Lindley, in Bot. Reg.*, 1840, *misc.* 30. Flowers very small, bright yellow, with rich red spots.—Guiana.—(Fig. 130, *flower, four times the natural size.*)

260. ONCIDIUM KARWINSKII. *Sertum orchidaceum*, 25. Flowers large, bright yellow, barred with brown. Lip white at the end, deep violet at the base.—Oaxaca.—(Fig. 131, *a flower, quarter the natural size.*) This is given to show how the Oncids differ from Miltonias, which are distinguished by the absence of warts, plates, or crest of any kind upon the base of the lip. Nothing of the kind being present here, the species is now called Miltonia Karwinskii. *See Journal of Hort. Soc.*, iv. 83, where is a full-sized figure of the flower.

261. ONCIDIUM PUMILUM. *Loddiges' Bot. Cab.*, t. 1732. Flowers very small, yellow, marbled with brown.—Brazil.—(Fig. 132, *flower, four times the natural size.*)

262. ONCIDIUM HARTWEGII. *Lindley, in Plantæ Hartwegianæ*, p. 151. Flowers small, brownish yellow, apparently whole-coloured.—Peru.—(Fig. 133, *twice the natural size.*)

263. ONCIDIUM UNGUICULATUM. *Lindley, in Journ. of Hort. Soc.*, 1. 303. Flowers pale green, speckled with crimson, and a clear yellow lip.—Mexico.—(Fig. 134, *column and lip, half natural size.*)

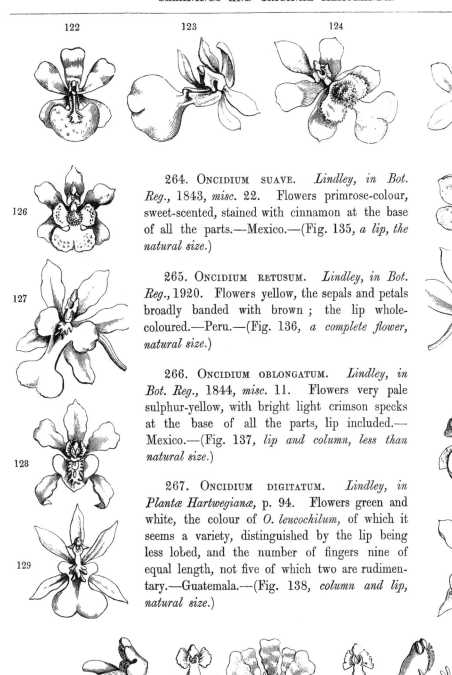

264. ONCIDIUM SUAVE. *Lindley, in Bot. Reg.,* 1843, *misc.* 22. Flowers primrose-colour, sweet-scented, stained with cinnamon at the base of all the parts.—Mexico.—(Fig. 135, *a lip, the natural size.*)

265. ONCIDIUM RETUSUM. *Lindley, in Bot. Reg.,* 1920. Flowers yellow, the sepals and petals broadly banded with brown; the lip whole-coloured.—Peru.—(Fig. 136, *a complete flower, natural size.*)

266. ONCIDIUM OBLONGATUM. *Lindley, in Bot. Reg.,* 1844, *misc.* 11. Flowers very pale sulphur-yellow, with bright light crimson specks at the base of all the parts, lip included.—Mexico.—(Fig. 137, *lip and column, less than natural size.*)

267. ONCIDIUM DIGITATUM. *Lindley, in Plantæ Hartwegianæ,* p. 94. Flowers green and white, the colour of *O. leucochilum,* of which it seems a variety, distinguished by the lip being less lobed, and the number of fingers nine of equal length, not five of which two are rudimentary.—Guatemala.—(Fig. 138, *column and lip, natural size.*)

268. PISTIA STRATIOTES. *Linnæus.* A hothouse floating plant of no beauty. Native of all parts of the Tropics. Flowers green, inconspicuous. Belongs to Lemnads (Pistiaceæ).

This now common, and, we think, very ugly plant, is thus spoken of by Sir William Hooker in the *Botanical Magazine*, t. 4564. We have only presumed to make a few indispensible corrections in the style :—

" With no floral beauty to recommend it, a more delicate and graceful object cannot well be seen in a tropical-house than tufts of *Pistia Stratiotes*, of the tenderest green imaginable, floating on the surface of a vessel of water or a tank. The leaves are connected together into a rose-shaped tuft, and these send out runners bearing other plants in all stages of growth. Dr. Roxburgh aptly compares them to half-grown lettuce plants. They continue in great beauty all summer and autumn, and in early winter they show symptoms of weakness or decay ; but, with a little care, plenty of young plants may be retained for the following spring, when they soon revive and reproduce by offsets. The inflorescence is nestled at the base of the leaf, and it may easily be seen there, by some of the young unfolded leaves, that the spathe which encloses the flowers is nothing but a modified leaf, the lower sides involute, and bearing the stamens and pistil. These flowers possess no beauty. The roots are a very pretty object when a plant is lifted out of the water, for here, as in the Duckweed (*Lemna*) of our own country—and *Pistia* is sometimes called tropical Duckweed,—the roots descend loose into the water, with no necessary attachment to soil or mud, and are long and feathery. Like many water plants it has a very extended range, perhaps all round the world, in tropical or subtropical regions. In America it extends as far north as Louisiana, the Mississippi and North Carolina. From Africa, I possess specimens from Egypt in the north, from the Niger country near the middle, and from Port Natal in the south. In the warm parts of India it seems to be universal. In Antigua, Patrick Browne tells us, it is most abundant in all the ponds of water preserved for public use ; that it keeps always fresh and cool the water, which would be greatly subject to putrefaction and charged with a multitude of insects, if it continued exposed to the heat of the sun. The plant, however, is there considered acrid, and when the droughts set in and the waters are reduced very low (which frequently happens in that island), they are overheated and so impregnated with the particles of this vegetable, that they occasion bloody fluxes to such as are obliged to use them at those seasons. I am aware that some botanists are disposed to consider that there are several distinct species of *Pistia*, and Professor Kunth goes so far as to constitute two groupes, and of one groupe to make two subgroupes, including altogether no less than *nine* species : but the characters are wretchedly defined, and I must confess, that as far as can be collected from the dried state of the copious specimens in my herbarium, there is no reason for constituting more than one species. Others, however, must judge for themselves. Our plant, here figured, is derived from Jamaica, and quite accords with Roxburgh's from the East Indies. Yet Sloane's Jamaica species (*Hist.* t. 2., f. 2.) is referred by Kunth to his *P. commutata*, and Brown's Jamaica plant to *P. obcordata.*"

Mr. Smith adds that :—" In this country it must be grown under glass, in a cistern or tank of water at a temperature ranging in summer between 70° and 80°. The depth of the water, whether several feet or only a few inches, is unimportant ; when it grows in deep water its roots do not reach the bottom. As it increases rapidly by producing stolons, or runners, in the form of rays, each of which bears a young plant, which becomes a new centre for producing stolons, it will, if allowed, soon occupy, in one summer, more space than can often be afforded for growing tropical aquatics. It will also grow freely in a small shallow tub or pan ; and, although its natural habit is to *float*, yet it appears to thrive more luxuriantly in water only a few inches deep, so that the roots reach the soil ; and it may be stimulated to grow to a size much larger than usual, by placing a thin layer of rich soil or very rotten dung in the vessel. Soft water is essential to its healthy cultivation, and in summer it should be shaded during the middle of the day, otherwise it is apt to become yellow and to have an unhealthy appearance."

269. RHODODENDRON MYRTIFOLIUM. *Schott.* A hardy evergreen shrub from the Alps of Southern Transylvania. Flowers red. Cultivated in the Garden of Schönbrunn.

This bush has been mistaken, according to Schott, for an Alpine form of Rhododendron ferrugineum. It is described thus :—Leaves minute, ovate or obovate elliptical, obtuse, rolled back at the edge and slightly crenelled, with a small point at the end; smooth, wrinkled, and dark green on the upper side, covered with a coarse shaggy wool on the under. Flowers in short racemes, about five together, with their stalks covered by a coarse scurf. Teeth of the calyx very short. Corolla funnel-shaped, the tube hairy outside, with some scattered scurf, the segments rounded, elliptical, smooth on each side near the rim, the throat shaggy, filaments hairy at the base, otherwise smooth. Style rather shorter than the ovary. Capsules lifted upon the lengthened peduncles above the leaves, and crowned by a short style.—*Botanische Zeitung*, 1851. 17.

270. JOSSINIA ORBICULATA. *De Candolle.* (*aliàs* Eugenia orbiculata *Lamarck ; aliàs* Myrtus orbiculata *Sprengel.*) A handsome evergreen stove shrub, from the Mauritius, with yellowish fragrant flowers, and firm flat oblong leaves. Introduced at Kew.

" A groupe of the *Myrtle* family, having very thick coriaceous leaves, axillary single-flowered peduncles, quaternary flowers, a bibracteolated calyx and numerous stamens, inhabiting Mauritius and the adjacent islands (Bourbon and Madagascar), called Bois de Nèfle (Medlar-wood), or Bois de Clou, on account of the hardness, by the colonists of

Mauritius, was formed into a genus (*Jossinia*) by Commerson, and adopted by De Candolle ; but by other botanists these plants are incorporated with *Myrtus*, and apparently justly so. The present species is from Mauritius, whence it was introduced into Kew Gardens, in the year 1824, and raised from seeds. Its flowering season is November, when its myrtle-like flowers, copiously nestled among the dark-green foliage, exhale the most delightful fragrance." The present species is " a stiff, branched, bushy shrub, seven feet high, and being a tropical plant, it requires the heat of the stove. It is a robust grower, thrives in any kind of light loam, and requires to be well supplied with water during summer. Being of a clean habit, and not subject to insects, it is suited for a select collection of stove-plants. It increases readily by cuttings, which should be planted in sand under a bell-glass, and plunged in bottom-heat."—*Bot. Mag.*, t. 4558.

We regret our inability to concur in the suppression of the genus Jossinia, which appears to form a very natural groupe of species, conspicuous for their hard broad leaves. If that genus is to be reduced to Myrtus, then the latter must be augmented by Eugenia, and the old farrago of species will be revived. Or if Jossinia is to disappear in any other genus, it must surely be in Eugenia and not in Myrtus. Dr. Wight assures us (*Illustrations of Indian Botany*, II., p. 12.) that the cotyledons of Jossinia are leafy, and neither cylindrical as in Myrtus, nor amygdaloid as in Eugenia. We must not, however, confound the genus Jossinia with Diospyros, as appears to be the case in gardens, as the next article will show.

271. DIOSPYROS AMPLEXICAULIS. A stove shrub, with hard dry alternate sessile leaves, which clasp the stem by their base. Native of the Mauritius. Flowers unknown. (Fig. 139.)

D. *amplexicaulis ;* glaber; foliis sessilibus amplexicaulibus coriaceis orbiculatis obtusis v. acutis, fructu turbinato ligneo 10-loculari 10-spermo in calyce coriaceo cupuliformi campanulato 6-lobo insidente.

An anonymous correspondent of the *Gardener's Chronicle* has sent us a leaf of what certainly is the plant now defined. He appears to have received it under the name of *Jossinia sp.*, and no doubt from the Mauritius : for we find it among dried plants of that island communicated by M. Bouton. How different it is from a Jossinia, whose fruit is fleshy and eatable like a medlar, will be seen by the accompanying figure taken from M. Bouton's specimen. It may also be added, that in the Jossinias the leaves are filled with transparent dots after the manner of their race ; while in Diospyros, as in all Ebenads, the leaves are dotless. It seems to be related to D. reticulata ; but its flowers being unknown this remains uncertain.

139

272. ASTER SIKKIMENSIS. *Hooker*. A handsome hardy perennial, with rich violet and yellow flowers. Native of the Sikkim Alps. Belongs to Composites. Introduced at Kew.

Raised from seeds sent by Dr. Hooker from the alpine regions of Sikkim. It flowers in October, and enlivens the garden at that late season with its copious bright purple flowers. It is remarkable of this and of Aster Cabulicus, that the stems form almost perfect wood the first year, three or four feet high, in the early winter abounding in leaf-buds, but dying down with our winter to the root. Stem erect, almost woody, and fragrant, three or four feet high, purplish-brown. Leaves glabrous, lanceolate, narrowly acuminated, spinuloso-serrate, with several parallel, very oblique nerves and numerous lesser connecting ones. Corymbs large, leafy, with numerous heads, which are purple. In the open ground it has every appearance of assuming the character of a hardy perennial.—*Bot. Mag.*, t. 4557.

273. CEREUS SUPERBUS. *Ehrenberg*. A large, dull red-flowered plant, from Mexico, with a weak branching seven-angled stem. In the Botanical Garden, Breslau.

This is said to have the appearance of some hybrid from *C. speciosissimus*, bearing larger flowers, of a light dull cinnabar-red colour, with carmine-red stamens. The stem is weak, branched, with club-shaped divisions contracted at the base, with seven angles, purple when young ; the ribs acute, crenated, with convex white downy cushions, bearing some shining nearly equal yellow prickles.—*Allgem. Gartenzeit*, 1850, p. 233.

274. PITCAIRNIA CINNABARINA. *Dietrich*. A fine stove Bromeliad, with spikes of brilliant red flowers. From Brazil. Introduced by Ohlendorff & Son, of Hamburgh.

The leaves of this species are quite entire, smooth, and reddish underneath. The racemes are about six inches long, one-sided ; the flowers quite smooth, about two inches long, of a deep rich vermilion red colour. Seems to be a very handsome plant.—*Allgem. Gartenzeit*, 1850, p. 202.

275. PIMELEA MACROCEPHALA. *Hooker.* A stiff glaucous greenhouse shrub, with large nodding heads of cream-coloured flowers. Native of Swan River. Belongs to Daphnads. Introduced by Lucombe & Co. (Fig. 140.)

140

One of the many Swan River plants raised from seeds received from Mr. Drummond. Perhaps its nearest affinity is with *P. tinctoria* Meisn., though the leaves do not change to the very peculiar green described as characteristic of that species, and it wants several other distinguishing marks. It is a highly interesting addition to our greenhouse plants, easy of culture, and free to blossom in the summer months. Shrub two to three feet high, somewhat simple, or fastigiately branched ; branches erect, smooth, rather robust (for a *Pimelea*), reddish below, green above, terete, leafy up to the involucre. Leaves opposite, smooth, the upper ones, especially, erect and secund, all of them large for the genus, and thick, rather leathery, broad-lanceolate, glaucous, acute, sessile; lower ones more spreading. Involucre of four to six leaves, larger and broader than the stem ones, shorter than the flower-head. This latter is two inches and a half across. Flowers numerous, dense, very pale rose (cream) colour. Calyx-tube slender, long, downy, articulated on the truncated summit of the ovary ; the segments oblong, spreading or recurved, ciliated at the margins. Stamens and style much exserted. Anthers orange.

"An Australian genus consisting of slender twiggy shrubs, and now numbering above fifty described species. The greater number are natives of Van Diemen's Land and the extra-tropical coasts of Australia, many being found at Swan River and at King George's Sound on the south-west coast : a few extend northward to within the tropics, and several are natives of New Zealand. About twenty species are known to have been introduced into the gardens of this country. The first was *P. linifolia* in 1793, followed by *P. rosea* in 1800 ; between the latter year and 1823, *P. drupacea* and *P. pauciflora* were introduced : the first two, being pretty flowering species, were frequent inmates in the greenhouse, whereas the two latter, having inconspicuous flowers, were seldom seen, except in collections where rarity and number of species were desired. In 1823 we were so fortunate as to raise plants of *P. decussata*, which, on account of its being of neat habit and a free and showy flowering species, soon became a favourite with cultivators, but has of late been in some measure superseded by its more showy rival, *P. spectabilis*, which was introduced about ten years ago. The species now figured is of recent introduction, and, from what we know of it, will turn out to be another showy species. It is, like its allies, a greenhouse plant, and grows vigorously if planted in turfy peat-soil, containing a little loam, and kept sufficiently drained. Over-watering is undesirable, especially during dull damp weather in winter and spring ; and in hot weather the sides of the pot must not be exposed to the direct rays of the sun. It will propagate by cuttings, placed under a bell glass, and treated in the usual way, but it has been found to produce the best plants if grafted on stocks of *P. decussata*."—*Bot. Mag.*, t. 4543.

276. IONOPSIS TENERA. *Lindley.* A very pretty stove Orchid from the West Indies, with panicles of delicate white or pale lilac flowers. (Fig. 141, *a*, a diminished figure ; *b*, flowers little more than natural size ; *c*, lip magnified.)

This seems to be common upon trees in many parts of the West Indies. It was first brought to notice by Sir Charles Lemon, who received it from the Havannah. It has recently been beautifully flowered by Messrs. Henderson of Pine-apple Place, who produced it at a meeting of the Horticultural Society last December ; having obtained it from Jamaica. It also occurs among Mr. Linden's dried plants, No. 484, from the Caraccas, where it was found by Fruck and Schlim. It is not impossible that it may be no more than a large-lipped variety of *I. utricularioides*, a plant we have long lost sight of ; but, until that can be certainly ascertained, the name should not be disturbed. If it shall turn out that these must be regarded as one species, then we suspect that *Ionopsis pallidiflora* will also have to be abandoned. What must be done, however, before any good opinion upon such a point can be given, will be to ascertain the exact state of the lip in all these plants. At fig. *c*, is a correct representation of what exists in the species now figured from Jamaica, and in those from Havannah, and the Caraccas, viz. : two small yellow ears stand at the very narrow rounded base of the lip ;

Ion. pallidiflora seems to have the same structure. The same thing is found in the widely different *Ion. paniculata*, and in *I. Gardneri*, another Brazilian species (No. 5875 of *Gardner's Herbarium*). In *I. pulchella* there are also two ears, but they stand on the lip far in advance of the very small unguis. On the other hand in *I. zonalis*, a very remarkable plant, with a band of deep violet at the base of the lip, flowered in 1848, by Mr. Alderman Copeland, there are four such ears.

277. WAHLENBERGIA VINCÆFLORA. *Decaisne.* (*aliàs* Campanula vincæflora *Ventenat.*) A hardy annual (or perennial), with rich azure blue, white-eyed flowers. Belongs to the Bellworts. Native of New Holland. (Fig. 142.)

This was originally introduced by the French many years ago, and published by Ventenat in his account of the plants at Malmaison. There it was lost. It has now been recovered, and promises to become a fine decoration

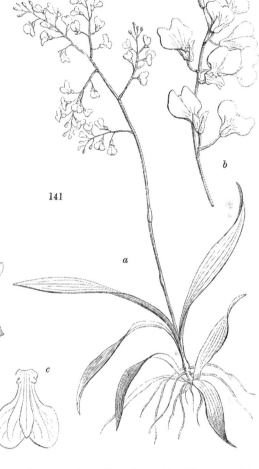

for our garden. Properly speaking it is a perennial ; but if so treated, its roots must be taken up and kept in a greenhouse in winter, for it cannot bear frost. It is, however, a very nice hardy annual, flowering about six weeks after being sown, provided it is put into a warm light soil, and the seeds are scarcely covered. The flowers are very pale on the outside, bright azure blue inside, furnished near the middle, and at the base, with a line of very delicate white hairs ; the tube is yellowish.—See *Revue Horticol.*, III., 41, where it is described by M. Decaisne.

278. SCHŒNIA OPPOSITIFOLIA. *Steetz.* A very pretty, hardy, annual from Swan River, with bright rose-coloured flower-heads. Belongs to Composites. Introduced at Kew in 1846.

" A lovely Swan River annual, quite equal in beauty to the *Lawrencella rosea* and to the *Rhodanthe Manglesii* of the same colony. Seeds were sent to us by Mr. Drummond, and our earliest plants blossomed in April, 1846. The genus is founded by Steetz ; and is nearly allied to *Helichrysum, Helipterum,* and still more to *Pteropogon* of De Candolle, from which it is said to differ by the inner scales of the involucre being appendaged and radiant, by the many-flowered capitula, and by the central florets being truly male.   The generic name is given in compliment to Dr. Schœn, an excellent botanical artist."

"This plant must be treated as a tender annual.   Its seed should be sown in spring, in a pot or pan of light soil, placed in moderate heat; the plants, as soon as they are of sufficient size, must be transplanted singly into small pots, and kept for a time in a close frame, admitting air gradually to harden them; and as they become larger they must be shifted into larger pots, and, in order to have a greater show of flowers, four or five plants may be placed in one pot. When in flower they may be placed in the greenhouse."—*Bot. Mag.*, t. 4560.

It is remarkable that an annual, of such beauty as this possesses, should not have become commonly known, after having been introduced at Kew for nearly five years.

279. CYMBIDIUM GIGANTEUM.   *Wallich.*   A noble terrestrial Orchid, native of Nepal.   Flowers deep warm yellow, with a richly spotted brown lip.   (Fig. 143, *a flower, natural size.*)

This is one of the most stately of Indian Orchids, producing strong, stiff, sword-shaped leaves in two rows, which of themselves render the plant a noble object in the stove.   The flowers add to this by their rich warm colours and large size. They appear at the end of a drooping scape about a foot and a half long, and covered with loose scales towards the base. Each is of the size represented in the accompanying cut, but turned upside down.   The sepals and petals are of a clear rich orange yellow; while the lip, which is bearded in the middle and at the edge, is richly mottled with cinnamon-brown. Owing to some error of observation, upon bad-dried specimens, we formerly reported the anther not to be articulated with the column as is usual, (see *Gen. & Sp. Orch.* p. 163,) but the fresh specimens show that the structure differs in no particular whatever from that of other genuine Cymbids.

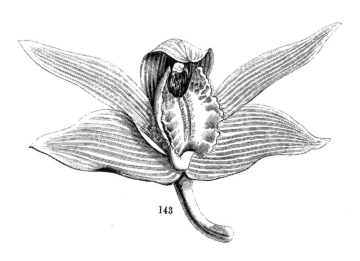

143

PLATE 40.

L. Constans, del & lith.

Printed by C.F.Cheffins, London.

[PLATE 40.]

THE FALSE SCARLET SAGE.

(SALVIA PSEUDOCOCCINEA.)

———◆———

A beautiful Greenhouse Herbaceous Plant, from S. AMERICA, *belonging to the Natural Order of* LABIATES.

Specific Character.

THE FALSE SCARLET SAGE.—Stem panicled, with spreading hairs, especially near the joints. Leaves stalked, somewhat heart-shaped, ovate, acute, crenated, glaucous or grey on the under side. Racemes simple. False whorls of 6 to 10 flowers each, as if leafless. Calyx narrow, striated, with the upper lip entire, the lower 3-toothed. Corolla obconical, downy, with the upper lip erect, undivided, the lower with three round divisions, of which that in the middle is broadest and emarginate. Stamens much projecting, with the abortive connectives linear, blunt, slightly adhering by the edges.

SALVIA *PSEUDOCOCCINEA ;* caule paniculato patentim piloso præsertim juxta nodos, foliis petiolatis subcordatis ovatis acutis crenatis subtus glaucescentibus, racemis simplicibus, verticillastris 6-10-floris subaphyllis, calycis angusti striati labio superiore integro inferiore 3-dentato, corollæ obconicæ pubescentis lab. superiore erecto indiviso inferioris laciniis rotundatis intermediâ majore emarginatâ, staminibus longè exsertis connectivis abortivis linearibus obtusis margine cohærentibus.

S. pseudococcinea : *Jacq. Icones, t.* 209. *Hooker in Bot. Mag. t.* 2864 (?) *fide Bentham, Genera et Sp. Lab.,* p. 290, *et De Cand. Prodr.* 12-343 : *aliàs* S. ciliata *Bentham, l. c.,* p. 286.

ONE of the finest of the gaudy species of Sage yet in cultivation is the present, which, although of old introduction, can scarcely be said to have been figured in any English work. It seems to be a common South American and Mexican plant, having been found there by almost all the collectors who have visited that part of the world. The specimen that furnished the accompanying plate was in the garden of the Horticultural Society, where it had been raised from seeds obtained in a London seed-shop, under the name of S. elegans.

It differs from *S. pulchella* in the form of its corolla, which is nearly exactly obconical, whereas in that species the corolla is very remarkably inflated on the lower side, below the contracted orifice. From *S. coccinea,* apparently lost in English gardens, it is said by Mr. Bentham to be distinguished by its much taller stem, covered with long spreading hairs (neglected by our artist) not hoary with short down, by its larger leaves, and by the upper lip of the corolla being rather longer.

Mr. Bentham quotes, as a synonym without doubt, the plate 2864 of the " Botanical Magazine," where, under the name of *Salvia pseudococcinea,* is figured a Trinidad plant obtained from the late Baron de Schach.   If this was really taken from a specimen of this species, it must be confessed that it conveys no idea of it ; for the blunt heart-shaped leaves, thin inflorescence, and shaggy flowers beyond which the stamens hardly project, are altogether at variance with *S. pseudococcinea,* as is the representation of the abortive connective of the anthers.   We suspect that the figure was really taken from a small specimen of *Salvia elegans* of Vahl.

The stem of this plant grows about three or four feet high, in a conservatory, and forms a large branched bush ; the sides, near the setting on of the leaves, are dotted with long white hairs.   The leaves have a rich deep green colour, grey on the underside, and are always drawn to a point : we have never found them blunt.   The flowers appear in great profusion, at the end of the branches, in the cheerless month of November, when their brilliant colour renders them most welcome.   Their corolla is somewhat curved, but if straight would present the figure of a nearly true cone, as far as the divisions of the limb : it is covered with close soft hair, not observable without a magnifying glass.   The abortive connectives of the anthers are rather shorter than the other half, linear, obtuse, a little inclined to hook at the point, pink, and adhere slightly by a few minute hairs produced on their contiguous edges.

PLATE 41.

L. Constans, Pinx & Zinc.

Printed by C.F. Cheffins, London.

[PLATE 41.]

THE BARBADOS CHERRY.

(MALPIGHIA GLABRA.)

———◆———

A Stove Shrub, from THE WEST INDIES, *belonging to the Natural Order of* MALPIGHIADS.

═══════════════

𝕾𝖕𝖊𝖈𝖎𝖋𝖎𝖈 𝕮𝖍𝖆𝖗𝖆𝖈𝖙𝖊𝖗.

THE BARBADOS CHERRY.—Young twigs and pedicels very slightly downy. Leaves lanceolate, acuminate, smooth when old, covered with acicular peltate hairs on the under side when quite young. Flowers in axillary stalked umbels. Glands of sepals 2 × 2 1 × 2 and 0. Petals fringed.

MALPIGHIA *GLABRA :* ramulis pedicellisque pubescentibus, foliis lanceolatis acuminatis vetustis glaberrimis novellis subtus pilis malpighiaceis tectis, umbellis axillaribus pedunculatis, sepalorum glandulis 2 × 2 1 × 2 et 0.

Malpighia glabra : *Linnæi Species Plantarum,* 609. *P. Browne, History of Jamaica,* p. 230. *Sloane's History of Jamaica,* vol. ii., p. 106, t. 207, f. 2. *Miller Ic.,* t. 181, f. 2. *Cavanilles, Dissert.* viii., t. 234, f. 1. *Adr. Juss. Malpighiac.* p. 11. *Ach. Rich. Fl. Cub.* p. 273.

═══════════════

WHOEVER has visited the West India Islands must have occasionally seen what are there called Cherries,—small succulent red rather angular fruits,—also called Chereezes, Brins d'Amour, and Gereceros. In Barbados they are especially plentiful, and have given its name to Cherry Hill, a well-known place in that colony. These fruits are the produce of the plant now figured, and of another called the Pomegranate-leaved (*M. punicifolia*); small trees from twenty to thirty feet high, long ago introduced to our gardens, but rarely seen in cultivation.

The plant from which our figure was taken was raised from Mexican seeds sent home by Hartweg to the Horticultural Society, in whose garden it flowers in September. It first began to blossom in 1847, and has done so every year since that time; but its flowers have never set, and English-grown fruit is still a desideratum. It is however a very pretty shrub, gay with its bright fringed rose-coloured blossoms, growing in little umbels from the axil of most of the leaves. It is not however a plant for exhibition-rooms, because the petals readily drop when the plant is shaken.

D 2

As we have already stated, two different plants bear the name of " Barbados Cherry," namely *Malpighia glabra* and *M. punicifolia*. They are however so very nearly allied, that their essential distinctions are not quite clear. Those which are pointed out by M. Adrien de Jussieu amount to little more than this; in *punicifolia* the leaves are " usually " notched at the point, and in *glabra* they are taper-pointed; in *glabra* the flowers are in umbels, the leaves covered beneath when young with those curious spicular double-pointed hairs attached by the middle, to which Botanists give the name of Malpighiaceous, and the innermost petal larger than the others (we find it about twice as large); while in *punicifolia* the flowers are solitary and the inner petal smallest. Nevertheless, Botanists seem to find it hard to distinguish the two.

According to Sir R. Schomburgk, *M. glabra* is called in Barbados the Red Cherry Tree. He describes the fruit of both species as being much used in preserves and tarts; but much inferior to the European Cherry; " there is something in the taste which reminds rather of the raspberry than the cherry." Ramon de la Sagra adds that the fruit becomes blackish when quite ripe. In Cuba he says that after subjecting them to boiling water they make with sugar an excellent preserve.

PLATE 42

L. Constans, Pinx & Zinc

1

2

3

Printed by C.F. Cheffins, London.

[PLATE 42.]

THE THREE-COLOURED VANDA.

(VANDA TRICOLOR.)

A beautiful Store Epiphyte, native of JAVA, *belonging to the Natural Order of* ORCHIDS.

Specific Character.

THE THREE-COLOURED VANDA. Leaves distichous, channelled, shorter than the few-flowered raceme. Sepals leathery, unguiculate, obovate, obtuse. Lip of the same length, three-lobed, with three lines in the axis. Spur short, obtuse; its lateral lobes rounded, broader than that in the middle, which is convex, cuneate and emarginate.

VANDA *TRICOLOR ;* foliis distichis canaliculatis racemo paucifloro longioribus, sepalis coriaceis unguiculatis obovatis obtusis, labello æquilongo trilobo per axin 3-lineato, calcare brevi obtuso, laciniis lateralibus rotundatis intermedio convexo cuneato emarginato latioribus.

Vanda tricolor : *Lindley in Bot. Reg.*, 1847, *sub t.* 59 ; *aliàs* V. suaveolens. *Blume Rumphia*, iv., p. 49, (1848).

THIS fine Orchid was first imported from Java by Messrs. Veitch, but has since reached England through other channels. It has the habit of *Vanda Roxburghii*, and its flowers appear in the same manner, but they are larger, have yellow and brown spotted sepals, and a rose-coloured lip, with the lateral lobes rounded, not acute, and colourless. It is near *V. Hindsii*, a New Guinea plant, not yet in cultivation; but that species has a long many-flowered raceme, extending as far as the points of the leaves. It has also been compared with *Vanda insignis*, an account of which has been published by Dr. Blume, with a figure, of a part of which the following is a copy :—

From this we learn that *Vanda insignis* has a concave, not convex, lip, with very small lateral lobes, and the broad central lobe deeply heart-shaped.

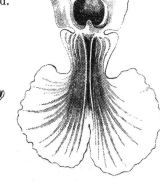

Many varieties of this species occur in collections, of which the three following are the most notable:—1. *V. t. pallens*. Flowers cream-coloured, with scattered brown spots. 2. *V. t. cinnamomea*. Flowers yellower, with lines of close cinnamon-coloured spots. 3. *V. t. planilabris*. With a clear citron ground-colour, scattered broad brown spots, and a *flat* purple lip. This looks very distinct from the others.

The figures at the bottom of this plate will serve to show how different *V. tricolor* is from *Vanda suavis*, fig. 3, and *Vanda Roxburghii*, fig. 2 ; of the latter the lip alone is given.

We avail ourselves of the present opportunity of giving a complete classified list of all the Vandas at present known.

AN ENUMERATION OF THE SPECIES OF VANDA.

SECT. A.—*Lip more or less lobed, divided, or expanded.*

1. Vanda teres *Lindl. in Wall. Cat.* no. 7324., *Bot. Reg.* t. 1809. ; *aliàs* Dendrobium teres *Wallich.*

V. foliis teretibus, racemis ascendentibus sub-bifloris foliis æqualibus, sepalis oblongis obtusis : supremo erecto lateralibus semitortis labello suppositis, petalis majoribus suborbiculatis undulatis, labello basi conico : laciniis lateralibus ascendentibus subtruncatis intermediâ pubescente apice dilatatâ truncatâ emarginatâ.

Native of hot damp jungles in *Sylhet, Burmah, Martaban,* scrambling up the bark of trees.

Flowers very large ; sepals white ; petals sanguine, with a white border. Lip sanguine, strongly veined, yellow on the upper surface below the point, and speckled with crimson. A most beautiful species.

2. Vanda cœrulea *Griffith MSS.* Plate 36 of this volume.

3. Vanda Roxburghii *R. Brown in Bot. Reg.,* t. 506. ; Vanda *Sir W. Jones ;* Cymbidium tesselloides *Roxb.*

V. caule brevi crasso, foliis apice obliquè tridentatis, racemis erectis foliis longioribus, sepalis petalisque oblongo-obovatis undulatis tessellatis obtusis, labelli lobo medio ovato emarginato obtusissimo canaliculato lateralibus acuminatis columnæ subæqualibus.

 var. α. sepalis petalisque sordidè luteis maculis obscurè fuscis, labello cœrulescente apice utrinque denticulato. —*Bot. Reg.* t. 506. *Fig. 2 in the annexed plate.*

 var. β. sepalis petalisque cupreis maculis oblongis luteis, labello lætè purpureo.—*Bot. Mag.* 2245.

Found on trees in many parts of the East Indies.

Flowers large, tessellated or whole-coloured, with a bright blue or purple lip. A beautiful species.

4. Vanda furva *Lindl. Gen. & Sp. Orch.* p. 215 ; *Blume Rumphia,* iv. 192, fig. 1, 197 c.; *aliàs* Angræcum furvum *Rumph.; aliàs* Epidendrum furvum *Linn.; aliàs* Cymbidium furvum *Willd.; aliàs*? Vanda fusco-viridis *Lindl. in Gard. Chron.* 1848, p. 351.

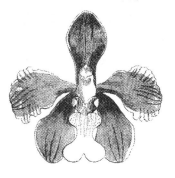

V. foliis canaliculatis rigidis apice obliquè retusis ; racemis erectis folio brevioribus laxis 3-5-floris, sepalis petalisque oblongo-obovatis curvatis, labelli lobis lateralibus ascendentibus obtusis intermedio patulo panduriformi apice rotundato emarginato plano lineis 5 elevatis, calcare obconico obtuso.—*Blume* quibusdam mutatis.

A native of the Moluccas.

According to Blume this has copper-coloured flowers with a pink lip. In *V. fusco-viridis,* which seems to be the same, they are described as dull brown, with a little greenish yellow at the edge, and a pure greenish yellow lip.

5. Vanda concolor *Blume Rumphia,* iv. p. 49 ; *aliàs* V. furva *Bot. Reg.,* aliàs V. Roxburghii unicolor *Hooker.*

V. caule alto, foliis laxis membranaceis apice obliquè tridentatis, racemis lateralibus plurifloris, floribus distantibus sepalis, petalisque oblongo-obovatis undulatis unicoloribus obtusis, labelli trilobi lobis lateralibus obtusis intermedio cuneato bilobo.

A native of China.

According to Blume this is not the *Angræcum furvum* of Rumphius, but a distinct species. It has the habit of *V. Roxburghii* but differs not only in the whole colour of its sepals and petals, and the other characters above indicated, but also in being a large lax-growing plant, five or six feet high, with much thinner and longer leaves.

6. Vanda limbata *Blume Rumphia,* iv. p. 49.

V. " labelli lobo medio arrecto panduriformi margine infernè revoluto apice rotundato-spathulato integerrimo."

Found in *Java.* (*Not in cultivation.*)

According to Blume the roots of this are very long. The flowers are ochre-coloured outside, brown and clouded on the inside; the lip lilac.

7. Vanda tricolor *of this plate.*

8. Vanda suavis *Lindl. in Gard. Chron.,* 1848, p. 351. *Fig. 3 of the annexed plate.*

V. racemis laxis brevibus, sepalis petalisque spathulatis retrorsis convexis valdè undulatis sublobatis apice rotundatis, labello convexo trilobo laciniâ mediâ altè bifidâ 3-costatâ lateralibus ovatis acutis patulis.

Reported to be a native of *Java.*

This has the foliage of *V. Roxburghii.* Flowers large, deliciously fragrant, white, with reddish-brown marbling and spotting. Lip deep violet. The lobed sepals and petals are remarkable ; they are both bent back at an angle of about 120°, and the petals are twisted round so as to present the principal part of their back to the eye.

9. Vanda Hindsii *Lindl. in Hook. Journ. Bot.*

V. foliis distichis arcuatis canaliculatis (pedalibus) apice obliquè emarginatis et excisis, racemo horizontali 10-floro foliorum longitudine, pedicellis floribus 3-plo longioribus, sepalis petalisque obovatis unguiculatis crispis, labelli cornu brevi obtuso lobo intermedio convexo cuneato apice rotundato: lateralibus abbreviatis rotundatis hinc acutis explanatis.

Found in the forests of *New Guinea* by the late Mr. Hinds. (*Not in cultivation.*)

This has the habit of *Vanda Roxburghii*, and its flowers seem to be of the same texture and size. Their colour cannot be judged of from the single dried specimen.

10. Vanda insignis *Blume Rumphia*, iv. p. 49. t. 192. fig. 2.

V. "foliis rigidis canaliculatis apice inæquali abscissis v. dentatis, racemis erectis folia adæquantibus laxis 5-7-floris, ph. perigon. obovato-oblongis rectiusculis, labelli lobis lateralibus ascendentibus obtusis intermedio arrecto apice dilatato rotundato undulato ad basin subhastatam e tuberculo obtuso cum lineis 2 elevatis, calcare obconico obtuso."—*Blume.*

Native of the mountains of *Timor.* (*Not in cultivation.*)

According to Blume's figure this would seem to be a most remarkable and handsome species. The flowers are 2½ inches . in diameter ; green outside, brownish inside. The lip is pink, white at the base, with a singular gauffered surface if we are to trust his figure 192, but flat as in other Vandas judging from his figure 197, reproduced on an adjoining page.

11. Vanda densiflora ; *aliàs* Saccolabium giganteum *Lindl. in Wall. Cat.* no. 7306.

V. foliis latis loratis carnosis apice uncinato-bilobis obliquis, racemis strictis cylindraceis multifloris, sepalis oblongis obtusis, petalis angustioribus obovatis, labelli calcare conico laminâ breviore obovatâ carnosâ apice 3-lobâ : lobis rotundatis intermedio inflexo minore.

A native of jungles in the *East Indies ;* (*Not in cultivation.*)

A reconsideration of the limits between *Vanda* and *Saccolabium* leads to the conclusion that this really belongs to the former genus, on account of its fleshy lobed lip and short spur. It has the habit of *Vanda multiflora.*

12. Vanda helvola *Blume Rumphia*, iv., p., 49.

V. "foliis rigidis subundulatis basi carinatis apice obliquè retusis, racemis erectiusculis folio brevioribus laxis sub-trifloris, ph. perig. oblongo-spathulatis lateralibus 2 exterioribus sub labello conniventibus, labelli saccati lobis lateralibus conniventibus obtusis intermedio patulo triangulari."—*Blume.*

Wild in mountain woods on the West of *Java*, flowering in March and April. (*Not in cultivation.*)

Blume states this to be a most magnificent species, forming a sort of transition between *Vanda, Renanthera* and *Cleisostoma.* Flowers the size of *V. suaveolens* (our *tricolor*), wine-red, shading into pale purple, with the lateral lobes a brighter purple.

13. Vanda longifolia *Lindl. in Wall. Cat.* no. 7322.

V. foliis longis loratis apice obliquis obtusissimis, racemis horizontalibus foliis triplò brevioribus, sepalis oblongis obtusis (undulatis ?), petalis angustioribus, labelli hypochilio concavo pubescente apice rotundato : cristâ carnosâ per axin, epichilio subrotundo-ovato obtuso.

A native of *Tavoy.* (*Not in cultivation.*)

Leaves a foot and more long. Flowers fleshy, apparently of the same size and character as in *V. multiflora.*

14. Vanda multiflora *Lindl. Collect. Bot.*, t. 33.

V. foliis loratis apice obliquis emarginatis, pedunculis sub-ramosis erectis subcorymbosis foliis brevioribus, sepalis petalisque oblongis obtusis maculatis subæqualibus, labello ecristato : lobo medio ovato acutiusculo basi lineâ mediâ pilosâ in calcar decurrente aucto.

Found wild in *China*, as well as in *Nepal.*

A species of no beauty, with coarse fleshy leaves and small yellow flowers dotted with sanguine.

15. Vanda congesta *Lindley in Bot. Reg. misc.*, 1839, no. 94 ; *aliàs* Saccolabium papillosam *Lindl. in Bot. Reg.* t. 1552 ; *aliàs* Thalia maravara *Rheede* ; Cymbidium præmorsum *Swartz.* ; Epidendrum præmorsum *Roxb.* Aerides undulatum *Smith.*

V. foliis ligulatis apice obliquis cuspidatis, racemis brevissimis capitatis, sepalis carnosis lineari-ovatis obtusis, labelli calcare obconico obtuso intùs villoso laminâ ovatá carnosâ papillosâ recurvâ.

Found in various parts of the *East Indies.*

A small-flowered species, with little axillary corymbs of yellow blossoms dotted with crimson.

16. Vanda parviflora *Lindl. in Bot. Reg.* 1844, *misc.* 57.

V. racemo simplici, sepalis oblongis, petalis linearibus spathulatis, labelli trilobi lobis lateralibus ascendentibus acutis intermedio oblongo canaliculato spongioso bilamellato apice circulari denticulato, calcare angusto obtuso.

Introduced from *Bombay* in 1843.

The flowers are small, pale ochre-coloured, with a lip sprinkled all over with extremely fine purple points ; the middle lobe of the lip is rather spongy, has two broad ridges, between which runs a channel, and at the point it is almost exactly circular, with a few small toothings.

17. Vanda spathulata *Spreng. Syst.* 3. 719. ; *aliàs* Ponnampou maravara *Rheede* ; Epidendrum spatulatum *Linn.* ; Limodorum spatulatum *Willd.* ; Aerides maculatum *Smith !*

V. foliis ovato-oblongis obtusis obliquè emarginatis, racemis erectis multifloris foliis et caule multò longioribus, sepalis petalisque oblongis obtusis planis, labelli hypochilio anticè bicalloso epichilio rhomboideo apice incurvo medio cristato, ovario hexaptero.

Native of *Mysore* and *Malabar.* (*Not in cultivation ?*)

A beautiful species with long corymbose racemes of golden-yellow flowers standing high above the short distichous leaves. Both leaves and flower-stalks are marked with crimson spots.

18. **Vanda lamellata** *Lindl. in Bot. Reg. misc.* 1838, no. 125.

V. foliis distichis coriaceis obliquè et acutè bidentatis, spicâ multiflorâ, sepalis petalisque obovatis obtusis undulatis inferioribus subincurvis majoribus, labello basi mammoso, limbo obcuneato retuso auriculato bilamellato ponè apicem bituberculato.

Found wild in the *Philippines :*

Flowers in long loose racemes, pale yellow, streaked with dull pale red. Not very handsome.

19. **Vanda peduncularis** *Lindl. Gen. & Sp. Orch.,* p. 216, no. 6.

V. foliis loratis apice altè et obliquè bilobis, pedunculo longissimo filiformi subramoso apice paucifloro corymboso, sepalis oblongis obtusis, petalis duplò minoribus, labello oblongo retuso villoso basi bilamellato.

Grows on trees in *Ceylon.* (*Not in cultivation.*)

Peduncle very long, slender, branched, purple. Flowers the size of *Ophrys apifera*, not unlike them. Sepals and petals pale green, streaked with purple. Lip shaggy, purple, bordered with green.

20. **Vanda cristata** *Lindl. Gen. & Sp.* no. 9 ; *Sertum Orchidaceum,* f. 3. in fronte ; *Bot. Reg.* t. 48.

V. foliis canaliculatis recurvis apice truncatis obliquè excisis tridentatis, racemo erecto trifloro foliis breviore, sepalis oblongis obtusis fornicatis, petalis angustioribus incurvis, labelli lobis lateralibus brevibus acutis intermedio vittato oblongo convexo apice saccato inæqualiter tricorni, cornu brevi conico.

Not uncommon in *Nepal.*

Flowers large, green, with a large broad lip, regularly striped with rich purple on a buff ground, and divided at the end into 2 narrow acute diverging lobes.

SECT. B.—*Lip contracted at the end, undivided, curved upwards or downwards.*

21. **Vanda Griffithii** (*Vandæ Sp. Griffith, Itinerary Notes,* p. 132, no. 846.)

V. facie foliisque V. cristatæ, floribus minoribus, labello basi concavo conico laciniis lateralibus nanis erectis intermediâ linguiformi convexâ emarginatâ basi minutè bicallosâ.

Found in *Bootan,* on the Monass River banks, on trees, at an elevation of 2300 feet. — *Griffith.* (*Not in cultivation.*)

In appearance this resembles a small specimen of *V. cristata.* Flowers yellowish-brown inside, and somewhat tessellated. Lip lilac, with deep yellow stains near the base. Capsule said to be nearly a span long, with six wings.

22. **Vanda Batemanni** *Lindl. in B. R.,* 1846, t. 59 ; *aliàs* Fieldia lissochiloides *Gaudich.*

V. radicibus crassissimis, foliis distichis coriaceis obliquè emarginatis obtusis racemo laterali multifloro brevioribus, bracteis coriaceis cucullatis squamæformibus, floribus maximis planis coriaceis, alabastris globosis, sepalis obovato-cuneiformibus obtusis, labello triangulari basi saccato lobis ascendentibus acutis apice carnoso sulcato uncinato dente elevato in medio et cristâ brevi transversâ juxta basin.

Found wild in the *Moluccas, Philippines* &c., growing on trees near the coast.

A very large erect plant, with remarkably thick aerial roots, sword-shaped curved two-ranked hard leaves averaging two feet in length, and a still longer spike of some score of flowers, each full two inches and a half across, flat, leathery, and long enduring. But it is not alone for their size that these flowers are so especially worthy of notice. Their colour is indescribably beautiful. If you look them in the face, they are the richest golden yellow, spotted all over with crimson ; but when seen from behind, they are wholly a vivid purple, fading away at the edges into the violet of *Cereus speciosissimus.*

23. **Vanda gigantea** *Lindl. in Wall. Cat.* no. 7326 ; *aliàs* V. Lindleyana *Griffith MSS.*

V. foliis latè loratis apice obtusissimis emarginatis subæqualibus, racemis foliis duplò brevioribus, sepalis petalisque oblongo-obovatis obtusis, labello incurvo canaliculato basi cordato apice dolabriformi obtuso : callo conico in medio.

A native of *Moulmein,* and other parts of the *Burmese* empire. (*Not in cultivation.*)

Leaves very long and broad, tough and fleshy. Flowers of the size of *V. Roxburghii*, resupinate ; deep yellow with cinnamon brown blotches. Mr. Griffith found it in flower at Mergui, March 1, 1835.

24. **Vanda Lowei** *Lindl. in Gard. Chron.* 1847, p. 239.

V. (foliis coriaceis rigidis distichis) ; racemo longissimo pendulo flexuoso scabro-piloso, floribus maximis distantibus coriaceis, sepalis petalisque lanceolatis acuminatis valdè undulatis extus scabris, labello parvo ovato glabro cucullato acutissimo supra medium cornu refracto setàque sub apice aucto.

Native of the forests of *Borneo* on high trees in very damp places.

Flowers lemon-yellow, barred and blotched with bands and spots of the richest cinnamon, three inches in diameter, disposed in pendulous racemes 10—12 feet long.

280. ACACIA LINEATA. *A. Cunningham.* A greenhouse shrub, from New South Wales, with heads of yellow flowers. Belongs to Leguminous plants. (Fig. 144. A, a magnified leaf and stipules.)

There is a figure of this plant in the *Botanical Magazine,* t. 3346 ; but it represents it in a glandular state very different from this. We find it to be a dwarf greenhouse shrub, flowering in March, without glands, but with a grey loose hairiness. The false leaves, or phyllodes, are linear, obtuse, a little hooked at the point, with a single rib running along the middle, but much nearer the upper than the lower edge. There is a very slight trace of a glandular depression on the false leaves, a little above the base, neglected in our figure.

281. MEDINILLA JAVANENSIS. *Blume.* (*aliàs* Melastoma javanensis *Blume.*) A stove shrub, from Java. Flowers pink, in a small panicle. Belongs to Melastomads. Introduced by Messrs. Rollisson.

This, though correctly referred by Blume to *Medinilla,* has little of the beauty of *M. speciosa* and *M. magnifica,* and others of the genus ; but it forms a handsome shrub, with ample, five-nerved foliage. We are indebted to Messrs. Rollisson, of Tooting, for this plant, which they imported through their collector from Java, along with another species of the genus, *M. crassifolia,* which has flowered at the same time with this, viz., in December, 1850. A shrub, with acutely four-sided, smooth branches. Leaves large, sessile, elliptical-ovate, acute, entire, five-nerved, somewhat cordate at the base, the rib red at the setting on of the leaf ; the general colour dark green, pale, and slightly tinged with red beneath. Panicle terminal and lateral (Blume), small, compact. This plant, being a native of Java, and, like others of the genus, subepiphytal, requires to be grown in a moist stove. A mixture of light loam and sandy peat soil, or leaf-mould, suits it. It should be well drained, and, as it is not a strong-rooting plant, care must be taken not to over-pot it. It propagates freely from cuttings treated in the usual way.—*Bot. Mag.,* t. 4569.

A

144

282. SMILACINA AMŒNA. *Wendland.* A mere weed from Guatemala, with a stem three or four feet high, and small green flowers. Raised at Herrenhausen. Belongs to Lilyworts.

The mould of some Orchids from Guatemala produced this plant. Its root is thick, clear-green and knobby. The leaves are 6-9 inches long, by 1½-3 inches broad, 5-7-ribbed, dull green, and shining on the upper-side, glaucous on the under. The flowers are small, white, in compound panicles, their stalks becoming reddish with age.—*Allgem. Gartenzeit*, 1850, 137.

283. PLATANTHERA INCISA. *Lindl. Gen. and Sp. Orch.* 293; (*aliàs* Orchis incisa *Willd*; *aliàs* Habenaria incisa *Sprengel*). A hardy herbaceous Orchid from N. America, with cylindrical spikes of purple flowers. (Fig. 145; A, a magnified flower.)

One of the large race of terrestrial Orchids, furnished with tubercles for roots, of which N. America possesses many, representing in its forest grounds and prairies the Common Orchids of Europe. The present is one of the rarer species, with purple flowers, having the lip deeply divided into three lobes, each of which is gashed and slit at the edge. It is nearly allied to the more common *Pl. fimbriata*, the flowers of which are larger, and the petals themselves deeply fringed. The stem of this plant is from 1½ to 2 feet high, covered with leaves like those of the Male Orchis. (*O. mascula*) but not spotted. The flowers themselves are of a deep lilac colour; and the bracts are so narrow and short as not to be observable among the flowers. The specimen from which the accompanying figure is taken, was received in July 1847, from Mr. Joseph Ellis, gardener to Henry Wheal, Esq., of Norwood Hall, near Sheffield.

284. DOMBEYA VIBURNIFLORA. *Bojer.* A broad-leaved, white-flowered shrub, from Madagascar. Belongs to Byttneriads. Flowered at Kew. Of little interest.

A native of the Comorin Islands, near Madagascar, according to Professor Bojer, who introduced the tree thence to Mauritius, from which latter island we have received it at Kew. With us, confined in a tub, it has attained a height of 12—14 feet, so that its rather small white flowers are not very conspicuous objects. Its nearest affinity is with *D. palmata* Wall., Pl. Asiat. Rariores; but the latter has seven spreading lobes to the leaves, and larger flowers with broader petals. It flowers with us in February.—*Bot. Mag.*, t. 4568.

285. SARRACENIA PURPUREA. *Linnæus.* A swamp herbaceous plant with dull purple flowers, from the United States. Belongs to Sarraceniads. (Fig. 146.)

Under the first plate of Vol. I. of the present work are various remarks touching the economy and cultivation of the curious race to which this belongs. To what was then said we now add some remarks by Dr. Asa Gray in his beautiful work on the Genera of United States plants :—

"The pitcher or open tube of the leaves evidently belongs to the petiole, which is also simply winged or margined along the inner side; while the blade is represented by the hood, or rounded appendage at the apex, which cannot be called a lid, as it never closes the orifice, nor is it so much incurved as at all to cover it, except in two species. This proper lamina is rudimentary in Heliamphora, and very small in proportion to the ample orifice which extends some way down the inner side, and thence a *double* wing-like border extends to the base, appearing just as if the two margins of an infolded leaf were united by a seam, so as to leave the free edges outside. In Sarracenia this wing, or margin, is simple and entire. The pitchers, especially those of *S. purpurea*, are generally found partly filled with water and dead flies, with other small insects. Whether the water is secreted by the leaf itself, or caught from the rain, is still undetermined. The point might readily be ascertained by proper observations, made especially upon *S. psittacina*, the pitchers of which are so protected by the hood that the fluid they contain (if any) can hardly be supposed to have entered by the orifice.

That the water in the open secreted by the internal hairs, suppose, would appear from are empty, and that during those of the previous season, species very long and deli- which alone or principally

"But, however derived, flies and other insects, which adapted to catch and retain.

pitchers of *S. purpurea* is not as Dr. Lindley and Mr. Bentham the fact that the younger leaves the spring and summer, it is from which these hairs (in this cate) have mostly disappeared, are found to contain water. this water serves to drown the these leaves are admirably According to Elliot and others,

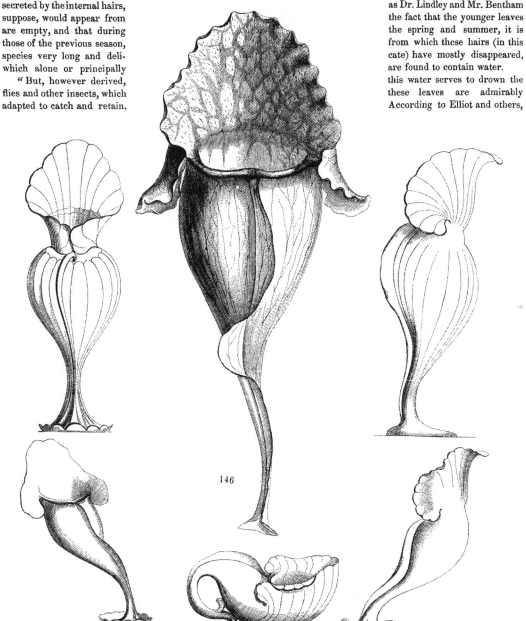

146

there is a saccharine exudation at the throat of the Southern species which attracts insects ; but this is not noticeable in *S. purpurea*. Immediately below the surface it is very smooth and polished, and still lower it is beset with sharp hairs, in most species long and slender, or else like those of the hood (in *S. Drummondii* extremely short and close), but in all pointing directly downward, so as to allow insects to descend, but effectually to obstruct their return. The inner surface of the hood is likewise lined with stiff and sharp retrorse bristles, which subserve a similar purpose, except in *S. flava*, which is smooth ; but in that species this appendage is erect with its sides turned away from the mouth of the tube, which thus it bears no part in guarding."

The species now represented inhabits the States from Virginia to Canada, and is readily known by the short inflated form of its pitchers. It strikes us that manufacturers might easily avail themselves of its grotesque figures for various economical purposes, especially the workers in gold and silver porcelain. In what way this may be done, our artist has endeavoured to suggest ; but we doubt not that the good taste and practical skill of manufacturers would soon strike out a better path.

286. SOBRALIA SESSILIS. *Lindley.* A stove terrestrial Orchid from Demerara, with solitary pale pink flowers of little beauty.

This was long ago well figured in the *Botanical Register*. Mr. Smith makes the following remarks upon the cultivation of the genus :—

" This is a species of a very pretty genus of terrestrial Orchids, natives of tropical America, growing in hot, dry places, and producing their showy flowers on the apex of slender reed-like stems, which rise from fascicles of thick, fleshy, interlacing roots. It requires to be kept in the warm division of the Orchid-house, and grows freely in a mixture of light loam and sandy peat. On account of its roots not going deep, it should be grown in a wide shallow pot, which must be well drained, so as to allow water to be given freely in summer without risk of the soil becoming saturated. It is increased by division of the roots ; but, in doing this, great caution is necessary, for, on account of their compact interlacing, they are not easily separated without injury."—*Bot. Mag.*, t. 4570.

287. RHYNCHOSPERMUM JASMINOIDES. *Lindley.* A Greenhouse evergreen climber, with white sweet-scented flowers. Native of China. Belongs to Dogbanes. (Fig. 147.)

This is a slender climbing evergreen shrub, rooting along its branches, whenever it touches a damp surface, like ivy. When wounded, its branches discharge a milky fluid. The young shoots are slightly downy ; the leaves opposite, oval, deep green, quite smooth, sharp-pointed, with minute scale-like glands in the place of stipules. The flowers are white,

147

deliciously sweet-scented, and produced in small irregular corymbs on the ends of peduncles, considerably larger than the leaves. Their calyx consists of five narrow smooth convex sepals, rolled backwards, and much shorter than the tube of the corolla, with a very shallow toothed glandular ring surrounding the base of the latter. The corolla is about three quarters of an inch long, pure white, salver-shaped, contracted in the middle of the tube, with a partially spreading border, whose five divisions are wedge-shaped, truncate, and twisted obliquely.

The anthers are five, arrow-headed, placed just within the orifice of the tube, and separated by five slightly elevated hairy lines. The ovary consists of two separate carpels, and is surrounded by five oblong green emarginate hypogynous scales, which sometimes are slightly united at the edge.

The structure of this plant is not precisely that of the genus Rhynchospermum, as given by M. Alph. De Candolle, for the scales beneath its ovary, are not exactly united into a cup. But they are partially so : and as there is no other difference as far as can be ascertained from the plant in a state of flowering only, it may be referred to the genus. In habit it is more like an Aganosma, but its corolla has not the tapering lobes of that genus, nor do the nectary or stigma correspond with it. *Journal of Hort. Soc.* vol. i, p. 74.

288. THIBAUDIA MACRANTHA. *Hooker.* A shrub of great beauty, from the jungles of India. Flowers drooping, very large, pale pink with blood-red veins. Belongs to the Cranberries (*Vacciniaceæ*). Introduced by Messrs. Veitch.

We represented what we considered to be the Prince of the East Indian *Thibaudias* in our Tab 4303 (*T. pulcherrima*), and in the rich abundance of its handsome flowers it has the superiority over this : but here, each individual flower is much larger and handsomer than in that species. We have measured these flowers two inches and a quarter long, and one inch in diameter ; the texture and marking resemble some handsome piece of china or porcelain. The plant was raised from seeds by Mr. Veitch, from Kola Mountain, Moulmain, whence they were sent by Mr. Thomas Lobb. It accords with many of the characters of *Thibaudia* (*Agapetes* De Cand.) *loranthifolia* Wall. ; but that species is downy, and differs in other points. We have rarely seen a more truly lovely plant. It flowered in the stove of Messrs. Veitch in December, 1850. Leaves on very short thick petioles, lanceolate, much acuminated, entire, glabrous. Flowers from the woody portion of the stem, extra-axillary. Two to three peduncles spring from the same point, and are pendent, thickened upwards, and red. Flowers large, and hanging down. Calyx small, pale yellow. Corolla large, pure china-white, yellow at the base and apex : the tube barrel-shaped, five-angled ; between the angles are numerous distinct, oblique, wavy red lines, generally taking the shape of the letter V, and more or less united : the mouth of the corolla is contracted : the five acute lobes reflexed. Stamens and style considerably exserted beyond the mouth of the corolla. We learn that it is an evergreen shrub of easy cultivation, and that it flowered when not more than two feet high. It is treated as a stove-plant ; but, judging from its allies and from its native climate, we are inclined to think it will succeed in a close greenhouse. Like many species of this family, the present is probably subepiphytal, deriving its chief nourishment from an atmosphere charged with moisture, and at a medium temperature ; such being the general character of the lower region of *Ericaceæ* and *Vacciniaceæ* within the tropics.—*Bot. Mag.*, t. 4566.

289. SAURAUJA MACROPHYLLA. *Linden.* A soft shaggy white-flowered stove shrub, from Guatemala. Belongs to Heathworts (Ericaceæ.) Blossoms in February. (Fig. 148 reduced ; A, the flowers the natural size.)

A correspondent in Edinburgh has sent us this through the post. He says that it came up among Orchids imported from Guatemala, by Mr. Skinner, and that it forms a vigorous shrub 3 or 4 feet high. The leaves are 6 to 8 inches long, covered with soft hair, obovate, tapering to the base, and serrated. The inflorescence is shaggy, with harsh hairs, panicled, with white flowers. It is evidently the *S. macrophylla* of Linden's Collections in the Caraccas (No. 106) : and must be nearly related to the *S. villosa* of De Candolle, for which we took it when it first reached us. It is not a showy species, but is useful among winter flowering things.

148

290. ECHINOPSIS CAMPYLACANTHA. *Pfeiffer.* (*aliàs* Echinocactus leucanthus *Gillies; aliàs* Cereus leucanthus *Pfeiffer.*) A long-spined Cactus, from the province of Mendoza, in the state of Chili. Flowers long-tubed, large, pink, with a grey outside. Produced at Kew.

A fine and well-marked species, with handsome flowers, readily distinguished by the great length of the central spine of the areolæ, and by its taking an upward and inward curve, a direction to which the other radiating spines are more or less inclined. It is a native of the Argentine province of Mendoza, at the eastern foot of the Andes, where it was discovered by the late Dr. Gillies, and introduced by him to our Gardens, with many others from that region, which we fear are now mostly lost to us. It flowers in the spring and summer months. Our *plants* are, the largest of them, a foot high, in shape between ovate and globose, not unlike that of a pine-apple, rather acute at the top, longitudinally furrowed ; *ridges* fourteen to sixteen, considerably elevated, scarcely compressed, obtuse ; the edges slightly tubercled or lobed.—*Bot. Mag.,* t. 4567.

291. SIPHOCAMPYLUS HAMATUS. *Wendland.* A stove shrub, covered with white down, and bearing violet flowers. Native of Brazil. Belongs to Lobeliads. In the Garden at Herrenhausen.

The stem is described as six feet high ; the leaves oblong-ovate or somewhat heart-shaped, with irregular callous teeth. The flowers in short dense racemes. Flower-stalks erect, as long as the corolla. Calyx with eight ribs and linear-lanceolate lobes, which are hooked backwards. The tube of the corolla narrow and slender. The whole plant is white with down.—*Allgem. Gartenzeit.* 1850, 138.

292. ECHINOCACTUS STREPTOCAULON. *Hooker.* A lumpish Cactus with numerous small yellow flowers, from Bolivia ; of mere botanical interest.

" A very distinct species of the genus *Echinocactus,* if we judge from the flowers ; but almost a *Cereus* in the elongated habit of the plant, which we purchased from Mr. Bridges, who had brought it from Bolivia. We find nothing like it anywhere described, and have named it from the remarkably spirally twisted character of the stem, without, however, holding ourselves responsible that this is a constant or permanent mark of distinction. It flowered in the Cactus-house of the Royal Gardens, in August 1845."

From some peculiarity in the nature of the Cactus region of Chili and Bolivia, we find that *Cacteæ* imported from these countries do not so readily conform themselves to the artificial modes of cultivation to which they are necessarily subjected in this country, as allied species from Mexico. This is more especially the case with the *Echinocacteæ.* We learn that they inhabit very arid and hot places, enduring extreme drought, which is very obvious from the harsh, dry, and often dead-like appearance they present when they arrive in this country. The species now figured was introduced with many others about six years ago, by Mr. Bridges, and on enquiring of him the nature of their places of growth, and what mode he would recommend as best for cultivating them in this country, the point on which he laid the greatest stress was to *give them no water.* But we find that even harsh, dry-looking *Cacteæ* are, like many other dry climate plants, capable of assuming a freer habit of growth by good treatment ; the difference of the growth they make in this country, as compared with that of their native country, is so great, that the top and lower part of the same plant, if separated, might be taken as two distinct species. It is probable that many *Cacteæ* from dry regions, when placed under the influence of a climate more favourable to vegetable development, will assume a different aspect, varying according to the degree of heat and moisture they receive.—*Bot. Mag.,* t. 4562.

293. TAMARINDUS INDICA. *Linnæus.* (*aliàs T. officinalis* Hooker.) A handsome tropical tree of the leguminous order, with pinnated small leaves, and racemes of yellowish flowers. Native of both Indies.

Most authors make two species of *Tamarindus,* the Indian kind with long pods, and the West Indian with short pods : but even those who adopt this view of the subject generally raise a question of their specific identity. India is probably the aboriginal country of both, whence the species was introduced to Western India. Even in the East the Tamarinds of the Archipelago are considered the best of those of India. The Arabs called the tree " *Tamar hindee,*" or *Indian Date,* from which has been derived the generic name *Tamarindus.* Our small Tamarind-tree, in the Royal Gardens, about fourteen feet high, whence our flowering specimens were taken, is probably the West Indian variety, and can give no idea of the general appearance of a full-grown tree, which all travellers agree in saying is one of the noblest objects in nature. " This most magnificent tree," says Dr. Roxburgh, " is one of the largest in India, with a most extensively spreading and shady head, or coma ; the bark dark-coloured and scabrous, the wood hard, very durable, and most beautifully veined." Dr. M'Fadyen, too, observes that the tree is " very ornamental, and affords a delightful shade." The inhabitants of the East, however, have a notion that it is dangerous to sleep under, and it has been remarked, as of our beech in Europe, that the ground beneath is always bare, and that no plant seems to thrive under its branches. Its flowers have little beauty to boast ; they are insignificant and exhibit no bright colours. Our plant has not borne fruit, but flowers in the summer season, and generally, but not always, casts its leaves during our winter. The extensive use of the pulpy fruits of the *Tamarind* is well known, as are its valuable medicinal properties. In the

East they are preserved without sugar, being merely dried in the sun, when they are exported from one part of the Archipelago to another, and cured in salt when sent to Europe. " In the West Indies," says the lamented Dr. M'Fadyen, " the pulp is usually packed in small kegs between layers of sugar, and hot syrup is poured on the whole. In order to enable them to keep without fermentation for a length of time, the first syrup, which is very acid, is poured off, and a second is added. A very excellent preserve is imported from Curaçoa, made from the unripe pods, preserved in sugar, with the addition of spices." The seeds are eaten in India in times of scarcity, by the poorer classes, the very astringent integument being first removed, and then, roasted or fried, are said to resemble the common field-bean in taste.— *Bot. Mag.*, t. 4563.

294. Passiflora penduliflora. *Bertero.* A green-flowered climber, of no beauty, from Jamaica. Blossoms at Kew in Spring and Summer.

Though destitute of the varied colouring of many of the species of the genus, there is a grace and elegance in this plant that render it an object well worthy of cultivation : the flowers are very copious, and hang downwards from

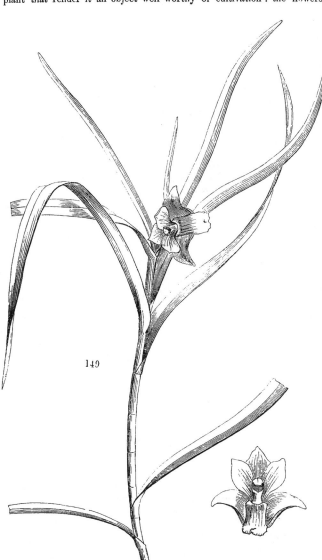

149

peduncles much longer than the leaves, and these leaves are very singular in shape. We received our plants from the island of Jamaica, where, indeed, it would appear to be very common. A climbing smooth shrub, with the young branches herbaceous and striated. Leaves numerous, close together, on very short petioles, varying a good deal in shape; but the general form is that of half an ellipsis approaching to cuneate, truncate, but more or less distinctly three-lobed, with three setæ, three-nerved, with a row of five or six glands on each side the midrib. Peduncles single-flowered, pendulous, jointed, and with two minute bracteoles above the base. Flower drooping, pale yellow-green. Calyx-tube hemispherical, ten-lobed : the five lobes of the limb oblong, very acute. Petals resembling the calyx-lobes, but a little longer. Nectariferous crown, deep orange, of from twelve to fourteen, short, nearly erect, club-shaped rays.—*Bot. Mag.*, t. 4565.

295. Ponera striata. *Lindl.* A grassy-leaved Epiphyte of no beauty. Flowers pale green. Belongs to Orchids. Native of Guatemala. (Fig. 149.)

This little-known plant has never been figured, although not uncommon in gardens. It has long grassy leaves upon a stem about two feet high; and, when old, throws out its flowers chiefly from the old leafless branches. The flowers are pale watery green, not unlike those of some Maxillariā, especially in the extended column-base with which the lip is jointed ; but the pollen-masses have the pulverulent caudicles of an Epidendrum. The lip is truncate as it were, and two-lobed, the one lobe over-lapping the, other. Two other species are known namely, *P. juncifolia*, with subulate leaves, from Mexico, not in Gardens, and *P. graminifolia*, also a Mexican plant, mentioned in the *Floral Cabinet* under the name of *Nemaconia graminifolia*.

296. ONCIDIUM BARBATUM. *Lindley*. A Brazilian Epiphyte; with panicles of small yellow and brown flowers. Belongs to Orchids. Blooms in January. (Fig. 150. a single flower, four times the natural size.)

　　Received from Parà by J. Knowles, Esq., of Manchester. It is evidently the little-known plant figured thirty years since in the Collectanea Botanica, and afterwards introduced by Mr. Gardner to the Glasgow Gardens, but apparently lost in the collections near London. It forms a small tuft of hard oblong one-ribbed pseudo-bulbs, having single oblong hard leaves much shorter than the branching stem. The flowers are yellow mottled with brown, and spotted with crimson on the lip. It differs from *O. ciliatum* in the petals being acuminate, not obtuse or emarginate, in the middle lobe of the lip being smaller (sometimes very much smaller) than the lateral lobes, and in the central tubercles of the crest being furnished with several smaller ones on each side, a circumstance overlooked in the figure by Mr. (now Sir William) Hooker, published in the Collectanea. The following woodcut is accurate in these important particulars.

150

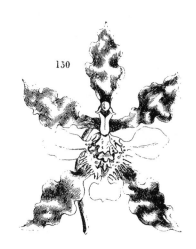

PLATE 4.

L. Constans. del. & Lith.

Printed by C.F. Cheffins, London

[PLATE 43.]

THE TWO-ROWED APONOGETE.

(APONOGETON DISTACHYON.)

———◆———

A hardy Aquatic, from the CAPE OF GOOD HOPE, *belonging to the Natural Order of* ARROWGRASSES
(*Juncaginaceæ*).

𝔖𝔭𝔢𝔠𝔦𝔣𝔦𝔠 𝔠𝔥𝔞𝔯𝔞𝔠𝔱𝔢𝔯.

THE TWO-ROWED APONOGETE. Leaves oblong-lanceolate, obtuse, seven-nerved ; spike two-parted ; bracts oblong, in two rows ; cap taper-pointed ; stamens twelve.	APONOGETON *DISTACHYON ;* foliis oblongo-lanceolatis obtusis septem-nerviis, spicâ bipartitâ bracteis oblongis distichis, calyptrâ acuminatà, staminibus 12.

Aponogeton distachyon : *Linnæi Supplementum,* p. 215 ; *Andrew's Repository,* t. 290 ; *Botanical Magazine,* t. 1292.

WE reproduce this plant in the hope of presenting a better figure of it than has yet appeared, and of drawing attention to a hardy aquatic of which too little is known. Several years have now elapsed since it was reported that a handsome sweet-scented water-plant from the Cape had been naturalised in the tanks of the Botanic Garden, Edinburgh. It was to the species before us that allusion was made, and it has since found its way, here and there, southward. It was introduced into Cornwall by Sir Charles Lemon, where, as well as in Devonshire, it seems to have as completely established itself as if it were a native of the county.

The correspondent, who first brought the Devonshire plant under our notice, expressed a doubt whether it was really the two-rowed Aponogete ; its flowers being so much larger than they are represented in books. There is, however, no doubt about its name being correct ; the differences that have been remarked being the mere result of exuberant luxuriance. The specimens came from an open pond at Woolston, the seat of the Rev. Charles Osmond, in the parish of Loddiswell, near Kingsbridge, S. Devon, where the plants thrive in a surprisingly luxuriant manner, producing thousands of delicious fragrant flowers throughout the summer ; and even in January bearing three hundred blossoms as fine as those represented. Innumerable seedling plants arise around their parents. To Mr. Osmond we are also indebted for the specimens, and for the following history of his acquisition of the plant :—

" About three years since a root was given me the size of a shot, which I planted in a small pan and sunk it in the pond; it grew rapidly, and, in a few months, produced flowers; and, unobserved by me, seed also, from which have sprung up to the surface of the water hundreds of plants. The spring which supplies the pond is peculiarly clear, always running, and, in the severest winter, rarely freezes."

The species is common at the Cape, where it bears the name of Water Uintjies. Mr. Bunbury mentions it thus :—

" The flowering tops of the Aponogeton distachyon, a pretty white-flowered floating plant, frequent in pools of water in various parts of the colony, are sometimes used both as a pickle and as a substitute for Asparagus."—*Residence at the Cape*, p. 208.

In appearance this resembles a Pondweed (*Potamogeton natans*), except that it is of a clear green colour without any tinge of brown. Its bulb (or corm) is described as being as large as a hen's egg. The leaves float on the surface of the water, are oblong, about 18 inches long when full grown, flat, and have three distinct veins running parallel with the main rib. When young their sides are rolled inwards. The flowers are placed on a forked inflorescence, originally included within a taper-pointed calyptrate spathe (cap), which is forced off as they advance in size. When fully formed each fork of the inflorescence is very pale green, and is bordered by two rows of large ovate-oblong obtuse ivory white bracts, in the axils of which stand the minute flowers. The latter are bisexual, and destitute of both calyx and corolla. Twelve hypogynous free stamens, with dark purple anthers, surround from four to six distinct carpels, each of which has a short curved style, a simple minute stigma, and six erect anatropal ovules. After flowering the bracts and inflorescence grow rapidly, acquire a deep green colour, and soon resemble tufts of leaves, among which lie in abundance large membranous green indehiscent beaked carpels, containing about four seeds each, and readily tearing at the sides. The seeds are exalbuminous, oblong, pale brown. The embryo is an oblong fleshy body, slit on one side, and in all respects is the same as in Triglochin; through the slit the plumule is pushed, while the seeds are still in their seed-vessels; germination beginning, apparently, as soon as the seeds come into contact with moisture.

These details sufficiently show that the natural affinity of the genus is with Potamogeton, Triglochin, &c., and by no means with the dicotyledonous Saururads, as Richard and others have imagined.

PLATE 44.

I. Constans. del & lith.

Printed by C.F. Cheffins, London.

[PLATE 44.]

THE SMALL-MOUTHED SIPHOCAMPYL.

(SIPHOCAMPYLUS MICROSTOMA.)

———◆———

A Stove half-climbing herbaceous Plant, from N. GRENADA, *belonging to the Natural Order of* LOBELIADS.

Specific Character.

THE SMALL-MOUTHED SIPHOCAMPYL.—Smooth, with a tendency to climb. Leaves stalked, ovate-oblong, with glandular serratures. Flower-stalks several times shorter than the leaf, naked. Tube of the calyx top-shaped, longer than the linear-lanceolate entire lobes. Corolla downy, straight, a little contracted at the mouth, with unequal lanceolate lobes. Anthers smooth, the two lower bristly at the point.

SIPHOCAMPYLUS *MICROSTOMA*; glaber, subscandens, foliis petiolatis ovato-oblongis glanduloso-serratis, pedicellis folio duplo triplove brevioribus nudis, tubo calycis turbinato lobis lineari-acuminatis integris tubo brevioribus, corollæ pubescentis rectæ ore paulò constricto lobis lanceolatis inæqualibus, antheris glabris 2 inferioribus apice setosis.

Siphocampylus microstoma : *Hooker in Botanical Magazine*, t. 4286.

OF this species Sir W. Hooker speaks in the following terms :—

" Among many fine species of Siphocampylus, detected by Mr. Purdie in New Grenada, few, if any, can vie with this, in the size of the flowers and richness of their colour. It seems also to produce its blossoms early and freely, and they continue a long time in perfection ; so much so, that though our plants were only raised from seed twelve months ago, they have been gay with flowers throughout the whole autumn and winter months, and have proved a great acquisition to our stoves, during this dreary season. In the summer, a greenhouse will be a better situation for it, and from the successions of buds that are forming, it seems to be one of those plants which one may reckon on having in bloom at all times of the year. Some of our plants have the stems and branches deeply tinged with purple, and the corollas are occasionally of a deeper and sometimes a paler scarlet, always produced in a compact leafy terminal umbel.

It is certainly a very pretty plant, and worth growing where stove plants can be well managed, especially since its glowing crimson flowers appear in midwinter; but it is apt to be the resort of the mealy bug, which seems to enjoy itself on its thin naked skin.

In most respects the species is extremely like the Peruvian Siphocampyl (*Syphocampylus Peruvianus*), from which, indeed, it is chiefly distinguished by its downy corolla and oblong leaves. With respect to the peculiarly contracted orifice of the corolla, to which it owes its specific name, either that is in some measure accidental, or the plants vary greatly with regard to it. In the specimen now figured it was scarcely observable; and we may add that while in the plant originally published by Sir William Hooker, the stamens were wholly enclosed within the corolla, as is often the case, in the plant which our artist had before him they were very fully protruded; so that such circumstances cannot be regarded as having any real value in the specific distinctions of this extensive genus.

L. Constans, del & Lith.

Printed by C.F.Cheffins, London.

[PLATE 45.]

THE TAPERING HOLLBÖLLIA.

(HOLLBÖLLIA ACUMINATA.)

———•———

A half-hardy evergreen Climber, from the NORTH OF INDIA, *belonging to the natural order of* LARDIZABALADS.

═══════════════

Specific Character.

THE TAPERING HOLLBOLLIA. Leaflets ternate and quinate, leathery, oblong-lanceolate, taper-pointed. Flower-stalks shorter than the leaf-stalks. Sepals very acute.

HOLLBÖLLIA *ACUMINATA ;* foliolis ternatis quinatisq. coriaceis oblongo-lanceolatis acuminatis, pedunculis petiolis brevioribus, sepalis acutissimis.

Hollböllia acuminata : *Lindley in the Journal of the Horticultural Society,* vol. ii., p. 4.

═══════════════

IN the mountain woods of Nepal grow two stout climbing shrubs named Hollböllia, by Dr. Wallich, after Mr. Frederick Louis Holboell, Superintendent of the Royal Botanic Garden at Copenhagen, "an experienced botanist, and a contributor to Hornemann's Flora Daniæ œconomica." Dr. Wallich also calls him "amicus et præceptor carissimus." They belong, with a few other plants, to a small natural order named Lardizabalads, the type of which is a Peruvian climber called *Lardizabala biternata.* Of these Hollböllias, the Broad-leaved and the Narrow-leaved, the first has been figured in the *Botanical Register,* for 1846, t. 49. The Narrow-leaved does not seem to be in cultivation. Dr. Wallich speaks of them as follows :—

"These two shrubs are easily distinguished from each other. The first species is by far the strongest, growing sometimes to a gigantic size. I brought specimens down with me, for the Honourable East India Company's museum, of a trunk, as thick as a good sized arm. Its leaves are broad, ovate, either ternate or quinate, about as long as the common petiole, the flowers quite white, collected in clusters ; the berries large and ovate ; the seeds oblong. The second species has long petioled leaves ; the leaflets from seven to nine, narrow, or linear-lanceolate, scarcely two-thirds of an inch broad ; the peduncles few flowered, and the flowers attaining, soon after expansion, a purplish colour ; the berries are not so thick, and of an oblong shape ; the seeds reniform. The natives of Nepal eat the fruit of both plants, the pulp of which has a sweetish, but otherwise insipid taste." Mr. Griffith found *Hollböllia latifolia* in woods near Churra and Moosmai in the Khasyah mountains, and confirms Dr. Wallich's account of the fruit, which he calls " Baccæ molles glabræ oblongæ albæ purpureo tinctæ.—*Itinerary Notes,* p. 36.

Possibly the plant now figured may have been regarded as a mere variety of the Broad-leaved, to which it no doubt approaches nearly. It seems, however, to be sufficiently distinguished by its long pointed leaves, which are very different from the blunt ones of *H. latifolia,* by its nearly sessile corymbs of flowers, and by its claret-coloured sharp-pointed sepals, the innermost of which are curved backwards at the point. Both species are fragrant; the species now figured resembled the orange flower in perfume. We find it among the dried specimens sent to Messrs. Veitch from the Khasyah Hills, by Mr. Thomas Lobb. It is thus mentioned in the Journal of the Horticultural Society :—

"This is an evergreen-twining plant, with quite the habit of *Hollböllia latifolia,* from which it differs in having very taper-pointed, not blunt, leaflets; racemes whose stalks are much shorter than the leaf-stalks, and purplish flowers not half the size; like that plant, it is deliciously fragrant. Hitherto male flowers only have been produced. It has been treated as a greenhouse climbing plant, but it is probably hardy; it grows freely in a mixture of sandy loam and rough peat, and is increased by cuttings. Its sweet-scented flowers, resembling the orange in perfume, and nearly evergreen foliage, make it a very desirable plant either in the greenhouse or open air."

GLEANINGS AND ORIGINAL MEMORANDA.

297. LYCASTE LEUCANTHA. *Klotzsch.* A pretty epiphyte, from Central America, belonging to Orchids. Flowers white, or stained with crimson. Introduced by M. Warczewitz. (Figs. 151, 152.)

There has recently appeared, among the plants obtained in Central America by M. Warczewitz, a species of Lycaste, remarkable for numerous varieties in the size and colour of the flowers. The first that blossomed, being quite white, received the provisional name of *L. candida,* now superseded by that of *leucantha,* published, in December last, in the *Allgemeine Gartenzeitung,* p. 402. During the present spring, others have appeared, with much smaller, and much larger flowers, some of which, instead of being colourless, are richly stained with crimson. Of the two accompanying figures 151 represents a large and colourless form ; 152, a smaller, blood-stained one. Among all these we find nothing like a

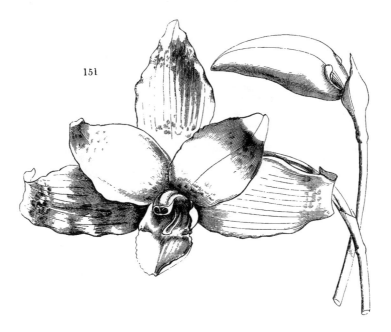

151

specific distinction, so that a description of one will apply very nearly to others. The species is among the dwarfest of the genus, its flowers scarcely reaching higher than the lower part of the leaves. A few distant sheathing-scales clothe the scape. The bract immediately beneath the flower is inflated in the middle, green, and contracted at the base, and as long as the ovary, which it nearly conceals. The flowers, by a curve of their slender stalk, arch over, so that, when fully expanded, they present their whole face to the eye of the observer. The sepals are oblong, rolled back at the point, somewhat wavy. The petals have the same form, but are smaller, and, when the flower is young, roll together at the lower half, so as to form a short tube (as in 152).   The lip is pure white, slightly 3-lobed, a little toothed at the edge, very concave, slightly hairy on the upper side, with a narrow, blunt, channelled appendage, lying along the middle, and not advancing more to the front than the separation of the obscure, round, lateral lobes of the lip. This column is, as usual in the genus, shaggy in front.

The species approaches *Lycaste plana* more nearly than any other, but differs not only in its much smaller size and more delicate habit, but in the middle lobe of the lip being concave, and slightly toothed, not convex, serrated, and plaited ; the appendage, too, is linear, not 3-toothed ; and the lip itself slightly, not deeply 3-lobed. It belongs to that part of the genus which is formed by *L. Skinneri, plana,* and their allies.

298. EUCALYPTUS GLOBULUS. *Labillardière.* A vast tree, from Van Diemen's Land, known under the name of the "Blue Gum."   Flowers white. Belongs to Myrtleblooms (*Myrtaceæ*). (Fig. 153.)

152

Two huge blocks of the timber of this tree having been sent from Van Diemen's Land by Sir William Denison, for exhibition in the "Crystal Palace," our readers will be glad to know something of its history. Garden catalogues say that it was introduced in 1810, and it is by no means rare among curious collections; but the rapidity of its growth soon renders it necessary to remove it.   There is, however, no reason why it should not thrive out of doors in the south west of England and Ireland, where the climate is as mild as in Van Diemen's Land.   It has angular branches which, when young, droop, and are of a pale dull green colour.   The leaves are firm, opaque, and unyielding, as if stamped out of horn, ovate-lanceolate, long-stalked, and curved in the form of a sickle; sometimes they are wider at the base on one side than on the other, and, by a twist of the stalk, always stand with their edges vertically instead of horizontally.   The white flowers are almost two inches across when the stamens are expanded ; and are produced singly or in clusters of threes ; sometimes, as in our figure, when the leaves fall off, the fruits seem as if in spikes.   The calyx is singularly knobby and rugged, with an angular tube, and a cover shaped like a depressed cone, or like a convexity with a rude boss in the centre. These flowers are covered before expansion with a thick glaucous bloom. The fruits are hard, woody, angular, rugged, knobby, urnshaped bodies, with five openings into the cavities of the capsule.

The early discoverers of this tree reported it to attain the height of 150 feet ; but they were far within the truth, as is shown by the blocks in the Great Exhibition, one of which near the base is 5 feet 7 inches in diameter ; and another, cut from 134 feet above the first, is still 2 feet 10 inches in diameter.   We learn from the proceedings of the Royal Society of Van Diemen's Land (vol. i., p. 157) that, on the 11th of October, 1848 :—

"A paper was read by Mr. H. Hull descriptive of a gigantic tree of the Gum tribe, ' occurring in a gorge on the declivity of the Mount Wellington range near Tolosa, about six miles from Hobart Town.' Mr. Hull describes it as a Blue Gum (*Eucalyptus globulus*), and says ' it stands close to the side of one of the small rivulets that issue from the mountain, and is surrounded with dense forest and underwood. * * * It was measured with a tape, and found to be twenty-eight yards in circumference at the ground (more than nine yards in diameter), and twenty-six yards in circumference at the height of six feet. The tree appeared sound except at one part, where the bark had opened, and showed a line of decayed wood. The full height of the tree is estimated to be 330 feet.' "

It is not improbable that the following extract from the same work (p. 165) relates to the same species, although it is spoken of by another name :—

"Mr. Milligan read the following note from the Rev. T. J. Ewing, of New Town, on the occurrence of some unprecedentedly large specimens of the *Swamp Gum* (*Eucalyptus Sp.*):—

"' NEW TOWN PARSONAGE, 19*th March*, 1849.

"' MY DEAR SIR,—I went last week to see a very large tree, or rather two very large ones, that I had heard of since 1841, but which were not re-discovered until Monday last. As they are two of the largest—if not *the* largest—trees ever measured, I have determined to send you an account of them, in order that a record may be preserved in any future publication of the Royal Society. They are within three quarters of a mile of each other, on a small stream, tributary to the north-west Bay River, pretty far up on the ridge which separates its waters from those of Brown's River. They are easily reached from the Huon foot-path, and are in a beautiful vale of sassafras and tree-ferns, and not in an inaccessible gully like most of our gigantic trees. I have never before seen the tree-ferns growing in such luxuriance, bending over the stream like enormous cornucopias. The fire has never reached them, as they and the forest around them plainly show ; and every here and there you are puzzled on seeing a sassafras tree with a root on either side—one in particular forming a natural arch, underneath which you can walk. And it was some time before I could tell how it was ever possible for the tree to have grown there, until, on looking further, I perceived that the sassafras must have originally sprung from seed lodged in the bark of some swamp gum that had fallen across the brook ; and, as it grew, it gradually sent out roots along the trunk until they met *terra firma.* The trunk having, in the course of ages, decayed, has left the sassafras tree in the odd position in which we now see it. I say so much before I give you the measurement. I am sure the whole scene would amply repay you for the trouble of a ride ; in addition to the giants below, there are, I feel confident, within a mile, at least a hundred trees of 40 feet in circumference. One, about forty yards from the biggest, was 60 feet at four feet from the ground, and, at a hundred and thirty, must have been fully 40 feet in circumference ; it was without buttresses, but went up one solid massive column, without the least symptom of decay. A silver wattle was 120 feet high, and 6 feet round. In fact, we named it the *Vale of Giants,* for puny indeed did men appear alongside these vegetable wonders. The largest we measured was, at three feet from the ground, 102 feet in circumference, and at the ground 130 feet. We had no means of estimating its height, so dense was the neighbouring forest, above which, however, it towered in majestic grandeur. This noble swamp gum is still growing, and shows no signs of decay ; it should be held sacred as the *largest* growing tree. The largest oak on record is the

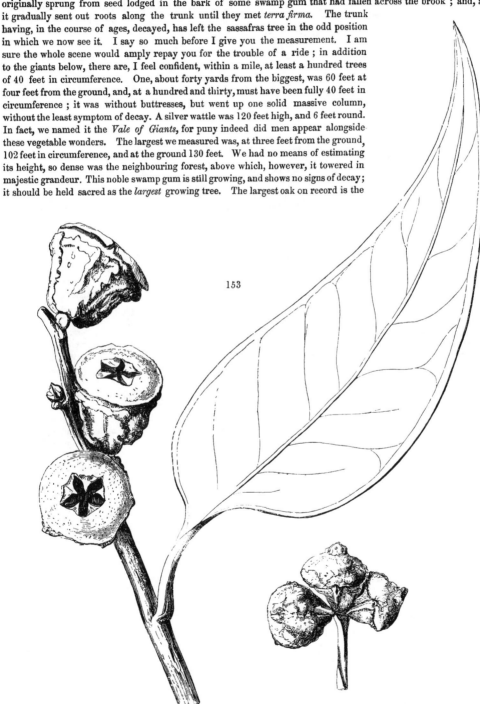

153

Cowthorpe, in Yorkshire, which is 48 feet in circumference at three feet from the ground. Some hollow pollard oaks are larger, such as the Winfarthing, in Norfolk, which is 70 feet at the ground. The second tree, also a swamp gum, is prostrate. It measures, from the root to the first branch, 220 feet, and the top measures 64—in all 284 feet, without including the small top, decayed and gone, which would carry it much beyond 300 feet. The circumference at the base is 36 feet, and at the first branch 12 feet, giving an average of 24 feet. This would allow for the solid bole, 10,120 feet of timber, without including any of the branches. Altogether, as green timber, it must have weighed more than 400 tons. The oak that gave the most timber was the Gelonos oak, in Monmouthshire, which, with its branches, turned out 2426 feet, but the body alone only 450 feet. * * * * .—Believe me, yours very truly,

"'THOMAS J. EWING.

"His Excellency the President mentioned his having strongly recommended to the Right Hon. the Secretary of State for the Colonies, and to the Lords Commissioners of the Admiralty, the timber of our Blue Gum (*Eucalyptus globulus*). Plank can be obtained from it in lengths surpassing those of any other timber tree ; and it may be sent home and sold at 8*d.* per foot, while oak plank (to which it is not inferior in quality), of the largest obtainable lengths, costs 2*s.* 6*d.* per foot."

Similar, although less striking, accounts of these gum trees are given by Mr. James Backhouse in his " Journal of a Visit to the Australian Colonies," as will be seen by the following extracts :—

" On an old road, called the Lopham Road, a few miles from the Bay, we measured some stringy bark (*Eucalyptus robusta*) trees, taking their circumference at about five feet from the ground. One of these, which was rather hollow at the bottom and broken at the top, was 49 feet round ; another that was solid, and supposed to be 200 feet high, was 41 feet round ; and a third, supposed to be 250 feet high, was 55½ feet round—as this tree spread much at the base, it would be nearly 70 feet in circumference at the surface of the ground. My companions spoke to each other when at the opposite side of this tree to myself, and *their voices sounded so distant, that I concluded they had inadvertently left me to see some other object*, and immediately called to them. They, in answer, remarked the distant sound of my voice, and inquired if I were behind the tree !" (P. 115.)

" In company with J. Milligan and Henry Stephenson, a servant of the company, from near Richmond in Yorkshire, we visited a place in the forest remarkable for an assemblage of gigantic stringy barks, and not far from the junction of the Emu River with the Loudwater, the latter of which takes its name from three falls over basaltic rock at short intervals, the highest of which is 17 feet. Within half a mile we measured standing trees as follows, at four feet from the ground. Several of them had one large excrescence at the base, and one or more far up the trunk :—No. 1, 45 feet in circumference, supposed height 180 feet ; the top was broken, as is the case with most large-trunked trees ; the trunk was a little injured by decay, but not hollow. This tree had an excrescence at the base, 12 feet across, and 6 feet high, protruding about 3 feet. No. 2, 37½ feet in circumference ; tubercled. No. 3, 35 feet in circumference ; distant from No. 2 about eighty yards. No. 4, 38 feet in circumference; distant from No. 3 about fifty yards. No. 5, 28 feet in circumference. No. 6, 30 feet in circumference. No. 7, 32 feet in circumference. No. 8, 55 feet in circumference ; supposed to be upwards of 200 feet high ; very little injured by decay ; it carried up its breadth much better than the large trees on the Lopham Road, and did not spread so much at the base. No. 9, 40½ feet in circumference ; sound and tall. No. 10, 48 feet in circumference ; tubercled, tall, with some cavities at the base, and much of the top gone.

"A prostrate tree near to No. 1, was 35 feet in circumference at the base, 22 feet at 66 feet up, 19 feet at 110 feet up ; there were two large branches at 120 feet ; the general head branched off at 150 feet ; the elevation of the tree, traceable by the branches on the ground, was 213 feet. *We ascended this tree on an inclined plane formed by one of its limbs and walked four abreast with ease upon its trunk !* In its fall it had overturned another, 168 feet high, which had brought up with its roots a ball of earth 20 feet across. It was so much imbedded in the earth that I could not get a string round it to measure its girth. This is often the case with fallen trees. On our return, I measured two stringy barks, near the houses at the Hampshire Hills, that had been felled for splitting into rails, each 180 feet long. Near to these, is a tree that has been felled, which is so large that it could not be cut into lengths for splitting, and a shed has been erected against it ; the *tree serving for the back !* (P. 121.)

As we have already observed, there seems to be no reason why these prodigious trees should not, at some future day, decorate the scenery of Great Britain. Devonshire and Cornwall, or Cork and Kerry, would certainly prove capable of bringing them to maturity.

299. HEBECLINIUM IANTHINUM. *Hooker.* (*aliàs* Conoclinium ianthinum *Morren.* *See* Vol. i., No. 172.)

Sir W. Hooker is of opinion that this plant should be referred to the genus *Hebeclinium*, rather than to *Conoclinium*, and that it is a close congener of *Hebeclinium macrophyllum*, a common plant of Jamaica, belonging to the first section of De Candolle. " As a species," he adds, the " plant differs abundantly in its large purple flowers and in the cuneate base to the leaf. It flowers in the winter months with us, and is then very ornamental. An herbaceous rather than a shrubby plant. Stem and branches terete, clothed with rusty down. Leaves opposite, on very long petioles, often a span long, ovate, but decidedly cuneate and entire at the base, very acute rather than acuminate, coarsely and often doubly serrated, the serratures mucronate. Corymb large, the capitula clustered at the ends of the branches. Flowers remarkable for the exceedingly long purple styles, which have, at first sight, almost the effect of a many-flowered ray. The corollas are also purple. Achenium angular. Pappus of few scabrous setæ."

If *Conoclinium* differs as a genus from *Hebeclinium*, merely in having a smooth conical receptacle, instead of a hairy convex one,—very small matters,—then no doubt this plant has been wrongly placed by Morren. But if the genera differ in the coloured enlarged bracts of the one, as compared with the herbaceous bracts of the other, then Morren's view may be the more correct. But, in truth, the genera are so nearly allied that it would be better to unite them than to waste words in unprofitable discussions concerning distinctions which are fleeting and undeterminate. Sir W. Hooker adds that the plant is Mexican and not Brazilian.

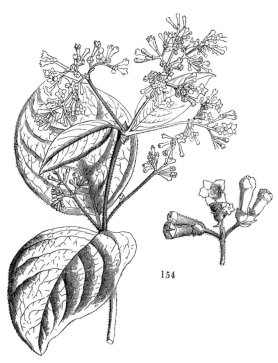

154

300. ROGIERA MENECHMA. *Planchon.* A stove shrub of the order of Cinchonads. Flowers pale salmon-coloured. Native of Guatemala. Introduced by the Horticultural Society. (Fig. 154, reduced, with flowers by the side, of the natural size.)

In his account of this genus, at t. 442 of the *Flore des Serres*, M. Planchon distinguished from his *R. amœna*, a plant which he called *R. Menechma*, by its stamens being inserted near the orifice of the tube, and having paler pollen, and by the shortness of the style, which does not reach half way up the corolla. In other respects it was said to be wonderfully like *R. amœna*. We entertain no doubt that his species is what is now figured from a specimen in one of the hothouses of the Horticultural Society, although in some respects the resemblance fails. It has the same manner of growth and similar foliage, but is more downy; and the leaves are more ovate. The branches are covered with a close fur instead of having a fine pubescence. The flowers are not so large nor so compactly arranged, and are much paler; the lobes of the corolla are almost acute instead of being emarginate; and the anthers are placed just below the throat of the corolla. We do not, however, find the style always as M. Planchon describes it; sometimes it is protruded beyond the orifice of the corolla as in *R. amœna*, sometimes it is not half the length of the corolla. Both these plants are very useful aids in decorating stoves, and possess the good quality of growing without unwillingness under the commonest management. Dampness, light, tropical warmth, and a light vegetable soil, are all the requisites which they demand.

301. TROPÆOLUM PENDULUM. *Klotzsch.* An annual (?) climbing, half-hardy plant, with yellow flowers, from Central America. Introduced by Mr. Mathieu, Nurseryman, Berlin.

Branches shining, round, bright green, climbing. Leaves peltate, smooth, glaucous beneath, deep green above, rounded and truncate at the base, slightly five-lobed, with short acute lobes of which the middle one is mucronate. Flowers axillary, solitary, pendulous. Calyx five-parted, yellow, with oblong lobes tapering to the point, the three upper curved backwards, the two lower nearly erect, together with the middle one of the upper set greenish at the point. Petals yellow, spathulate, crenated on the upper edge, the three lower long-stalked and whole coloured, the two upper sessile, recurved, marked with parallel red lines and a dull violet bar near the edge. Filaments yellowish. Anthers greyish green. Raised from M. Warczewitsch's collections. *Allgem. Gartenzeit.*, Nov. 30, 1850.

302. EPIDENDRUM ACICULARE. *Bateman, in Bot. Reg.*, 1841, *misc.* 98; *Supra*, vol. i., p. 150, *sub* t. 30 (*aliàs* E. linearifolium *Bot. Mag.*, t. 4572). A Mexican epiphyte, with rather gay panicles of purplish blossoms, and very narrow leaves. Introduced about 1840. Flowers in June.

This pretty species, to which Sir William Hooker has lately given the name of *linearifolium*, upon the supposition that it was undescribed, was originally published by Mr. Bateman, in 1841, as a native of the Bahamas, upon the authority of Mr. Skinner. We have seen no native specimens from that island; but it is undoubtedly Mexican, being

No. 30, of Galeotti's collection, gathered near Oaxaca, at an elevation of 3000 feet. We cannot, however, recognise it in Professor Achille Richard's brief enumeration of Galeotti's Orchids. It is doubtful whether the *E. bractescens* of the *Botanical Register* (1840), gathered by Hartweg near the same town, is really distinct notwithstanding its long leafy bracts. Both are closely related to the *E. microbulbon* of Hooker, accidentally omitted in the enumeration under t. 30 of our last volume, *E. ovulum, Linkianum,* and *Pastoris.* The species is admirably figured in the *Botanical Magazine,* with the following description :—

"Pseudo-bulbs scarcely exceeding an inch in length, clustered, ovate, quite even on the surface, the younger ones more or less sheathed with scales, bearing at their summit two very narrow linear leaves, eight to ten inches long, carinate, acute. The scape rises from between the two leaves, and is a foot long, bearing a lax slender graceful panicle of from twelve to fourteen flowers. Sepals and petals spreading horizontally, purple-brown, yellowish at the apex, very acute. Lip with its base united to the lower part and decurrent with the long column, the sides embracing and including the latter, three-lobed, yellowish white, delicately lined and veined with purple ; side lobes oblong, acute, reflexed ; middle lobe large, rotundate, waved at the margin. Column yellow, with blood-red spots, biaurite in front near the summit. Anther-case white, with crimson spots."

155

303. ACACIA RICEANA. *Henslow.* (*aliàs* A. setigera *Hooker.*) A tall drooping bush with long spikes of pale yellow flowers. Native of Van Diemen's Land. Flowers from February to May. (Fig. 155.)

The discovery of this most beautiful species was made by Mr. Ronald Gunn, who sent home dried specimens in 1837. In his unpublished notes he describes it as a very common species at Hobart Town, and on the banks of the Derwent ; but it was not seen by him on the north side of the colony. He adds, that it grows from six to ten feet high ; if planted out in the border of a greenhouse it grows much larger ; in the garden of the Horticultural Society it forms a bush of extreme elegance, rising twenty-five feet high, in the great iron conservatory, and then curving downwards its weeping branches, which are loaded with heaps of pale yellow flowers till the middle of April. When treated thus, *A. Riceana* is probably the handsomest species of its genus. The phyllodes, which grow in clusters, are linear, deep green, and sharpened into a fine point which itself is a continuation of a solitary rib which passes along the middle ; marginal gland there is none. The flowers grow singly in long loose spikes, and, before expansion, constitute small oval bodies with three short scale-like sepals and three petals.

The species was named by Professor Henslow after Lord Monteagle, then the Right Hon. T. Spring Rice, one of the members for Cambridge. It is readily distinguished from *A. juniperina* by the latter having its flowers in solitary spherical heads, not in long loose spikes.

304. ACACIA OXYCEDRUS. *Sieber.* (*aliàs* A. taxifolia *Loddiges.*) A handsome bush from the southern parts of Australia. Flowers in bright yellow spikes, appearing in January and February. (Fig. 156.)

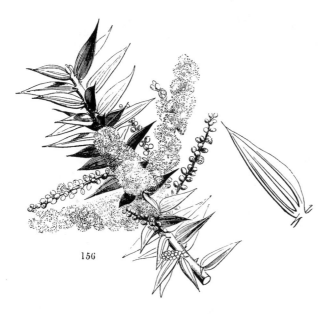

156

Apparently a common plant in Van Diemen's Land, and the south-eastern districts of the Australian continent. Sir Thomas Mitchell found it on Mount William in 1836 ; the Blue Mountains are named by others, and Mr. Backhouse notices it among the Tasmannian plants which struck him with their beauty. He speaks of it as being in flower on the 7th of September, among the earliest indications of spring, and again in April :—

"On the 15th of the fourth month," he says, " we held a meeting with some sawyers, in their huts, at a place called the King's Pits, on the ascent of Mount Wellington, at an elevation of about 2000 feet, and about four miles from the town. The forest among which they are residing is very lofty : many of the trees are clear of branches for upwards of 100 feet. It caught fire a few months ago, and some of the men narrowly escaped. The trees are blackened to the top, but are beginning to shoot again from their charred stems. The brushwood is very thick in some of these forests. A shower of snow fell while we were at the place. *Acacia oxycedrus*, ten feet high, was in flower on the ascent of the mountain. This, along with numerous shrubs of other kinds, formed impervious thickets in some places ; while, in others, *Epacris impressa* displayed its brilliant blossoms of crimson and of rose-colour."

In cultivation it forms a stout shrub, with hard, stiff, bright green phyllodes, having three strong ribs terminating in a fine point. In form these phyllodes are variable, sometimes being narrowly ovate-lanceolate, and somewhat falcate, or even linear, or so short and broad as to be almost ovate ; *A. mœsta* of the *Botanical Register* may even be a peculiarly broad variety. From *A. verticillata* the Oxycedrus is distinguished by its phyllodes having three or four distinct stout ribs, and not being whorled, its much stouter and more erect habit, and its larger and finer flower spikes.

305. ACACIA DIFFUSA. *Ker.* (*aliàs* A. prostrata *Loddiges.*) A handsome leguminous bush, from Van Diemen's Land, with numerous balls of bright yellow flowers appearing in midwinter. (Fig. 157.)

Although this has naturally a trailing mode of growth, yet it readily lends itself to the art of the gardener, and, by a little management, will assume the form of a close compact bush. It is extremely common in Van Diemen's Land ; varying greatly in the size and shape of the phyllodes (leaves), and in the length of the flower-stalks, which are sometimes nearly sessile, and sometimes on long stalks as in our figure. The phyllodes have a single rib, running from end to end, and terminating in a hard spine ; and, near their base,

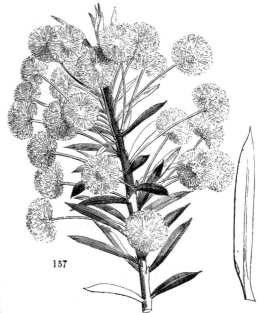

157

often appears a small oval gland, but this is frequently missing. When dry or old the phyllodes seem to have several veins lying irregularly by the side of the midrib ; but in reality this appearance is produced by the shrinking of the parenchyma, and the seeming veins are merely wrinkles. Although there is no difference whatever between the *Acacia diffusa* of Ker, and the plant afterwards published under the name of *prostrata*, in the *Botanical Cabinet*, yet we find specimens bearing in gardens both names, and not uncommonly with both names misapplied. The accompanying figure represents a piece of a very vigorous plant growing, as it should grow, when cultivated properly.

In Van Diemen's Land there is another species like this, and probably in our gardens, viz., the *A. siliculæformis* of Cunningham, a much smaller plant, with very smooth, almost shining, phyllodes, which never become wrinkled, and are not more than half the size of those of *A. diffusa*.

306. ACACIA UROPHYLLA. *Bentham.* (*aliàs* Acacia smilacifolia *Fielding, Sertum Plantarum,* t. 3. (1843).) A handsome greenhouse shrub, with pale yellow balls of flowers. Native of Swan River Colony. Flowers in January and February.

" Would that all the species of the vast groupe of phyllodineous *Acaciæ* were as easily defined as this. The phyllodia are here of a very peculiar character, generally broad ovate, subfalcate, almost spinescently acuminated, with longitudinal and transverse nerves, as in *Smilax*, whence the appropriate name of Mr. Fielding. The plant was raised from seeds, sent in 1843, by Mr. Drummond, from the Swan River Colony. It flowers in January and February."—*Bot. Mag.,* t. 4573. (According to Preiss among mud and stones in shady places, among the mountains continuing Darlings range, not far from the head-waters of the Swan ; and also in damp shady places on the river Canning, flowering in the cold season.)

It is described as a moderate-sized shrub, with angular branches, and the young phyllodes pubescent. Phyllodes obliquely ovate, slightly falcate, drawn into a slender setaceous point, hairy or glabrous, the upper edge obscurely crenate, the two surfaces marked with three nearly equidistant nerves, united by transverse ones, tapering at the base more or less gradually into a rather short footstalk, which bears a conspicuous gland at its summit above. Stipules two, minute, subulate, red, spinescent. Peduncles two to five from one axil, each much shorter than the leaf, longer than the petiole, each bearing a single head of pale yellow flowers, acute lobes.

307. QUERCUS AGRIFOLIA. *Née.* A hardy evergreen oak from California. Introduced by the Horticultural Society.

A few miserable living plants of this species were sent home by Hartweg from California, and are now beginning to grow in the Society's Garden. It will probably be a hardy evergreen tree, concerning which Nuttall, who knew it in its native country, has the following remarks :—

" This species, almost the only one which attains the magnitude of a tree in Upper California, is abundantly dispersed over the plain on which St. Barbara is situated, and, being evergreen, forms a conspicuous and predominant feature in the vegetation of this remote and singular part of the Western world. It appears more sparingly around Monterey, and scarcely extends on the north as far as the line of the Oregon territory. It attains the height of about 40 or 50 feet, with a diameter rarely exceeding 18 inches ; the bark is nearly as rough as in the Red Oak. The wood, hard and brittle and reddish, is used only for purposes of fuel, or the coarse construction of log-cabins.

" As an ornamental tree for the south of Europe or the warmer States of the Union, we may recommend this species. It forms a roundish summit, and spreads but little till it attains a considerable age. As a hedge it would form a very close shelter, and the leaves, evergreen and nearly as prickly as a holly, would render it almost impervious to most animals. The leaves vary from roundish ovate to elliptic, and are of a thick rigid consistence ; the serratures are quite sharp ; the young shoots are covered more or less with stellate hairs, and for some time tufts of this kind of down remain on the under side of the midrib of the leaves, which are, however, at length perfectly smooth, and of a dark-green above, often tinged with brownish yellow beneath. The staminiferous flowers are very abundant, and rather conspicuous; the racemes the length of 3 or 4 inches ; the flowers with a conspicuous calyx and 8 or 10 stamens ; the female or fruit-bearing flowers are usually in pairs in the axils, or juncture of the leaf with the stem, and sessile, or without stalks. The cup of the acorn is hemispherical, and furnished with loose brownish scales ; the acorn, much longer than the cup, is ovate and pointed. We do not recollect to have seen this tree properly associated with any other, except occasionally the *Platanus racemosa ;* their shade is hostile to almost every kind of undergrowth. By Persoon this species is said to have been found on the eastern coast of North America, while Pursh attributes it to the north-west coast, about Nootka Sound. It does not, however, extend even to the territory of Oregon, as far as my observation goes." Née says, " I have only seen branches collected at Monterey and Nootka. The leaves of the young plants are perfectly smooth when first developed, of a thin consistence, with numerous slender sharp dentures beneath ; they are of a brownish yellow colour, and appear smooth and shining." The long narrow acorns, almost conical, are a remarkable feature in the species.— *Journal of the Horticultural Society,* vol. vi., p. 157.

308. CHYSIS AUREA. *Lindley.* A stove epiphyte belonging to Orchids, with rich golden yellow flowers. Native of Equatorial America. Flowers in January.

From the collection of Messrs. Lucombe and Pince, of Exeter, by whom it was purchased at one of Mr. Stevens's sales of Columbian Orchideæ, in 1850, as the " Red Bull's-mouth." The specimen figured in the *Bot. Mag.* t. 4576, under the name of *Ch. aurea,* var. *maculata,* Sir W. Hooker was at first disposed to consider a distinct species from *C. aurea, lævis,* or *bractescens,* but a further investigation led him to the conclusion that it was rather a highly coloured variety of *C. aurea,* to which he observes that " *C. bractescens* is very nearly allied, nor do I find the chief distinction which

Dr. Lindley lays stress upon, available ; viz. that on the labellum of *C. aurea* there are five principal ridges, and three minor ones on each side, all downy and diverging, 'while in *C. bractescens* there are five equal ridges all smooth and parallel. In our drawing of *C. bractescens*, now before us, the five ridges are all downy in their lower half, while in *C. aurea*, both α and β, the three lesser lateral ridges appear rather a kind of venation, such as is seen in the middle lobe also. In *C. bractescens*, the bracteas are larger and very concave, and the flowers are larger, and the lateral lobes of the labellum are larger than in *C. aurea*. The flowers are very fragrant.' "

Upon again referring to the materials in our possession for illustrating the differences in the three species of *Chysis*, we find little to alter in what was formerly said about them. The principal ridges at the base of the lip of *Ch. bractescens* are, no doubt, downy half way up, as Sir W. Hooker states, and they vary in number from 5 to 7 ; but they are much blunter than in *Ch. aurea*, and the lip is wholly destitute, in our specimens, of the lateral hairy veins peculiar to *Ch. aurea*. The most material difference between these species is, however, the great inflated bracts of *Ch. bractescens*, to which there is no approach in *Ch. aurea*. As to *Ch. lævis* it has the bracts of the latter, from which it is distinguished by a shorter middle lobe of the lip and smooth short ridges, the two lateral of which are rudimentary. We have not seen it alive since July, 1840.

309. BERBERIS PALLIDA. *Bentham*. A beautiful evergreen greenhouse bush, from Mexico. Flowers yellow. Berries black. Flowers in the early spring ; fruits in autumn and winter. (Fig. 158.)

We learn from the *Botanical Register* that in its native country this forms an evergreen shrub from five to six feet high, and is found but sparingly, near

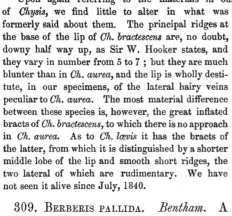

158

Cardonal and Zimapan, on mountains thinly covered with *Pinus Llaveana.* Hartweg also met with it near the hot springs of Atotonilco El Grande, but nowhere in any quantity. It is easily distinguished by its dry hard leaves and pale yellow flowers. The wood is also said to be of a lighter colour than in any other species. It grows freely when potted in a mixture of sandy loam and leaf-mould, to which is added a small portion of rough bone-dust. It may be increased like other pinnated berberries, by grafting on the common *B. Aquifolium,* either in spring or Autumn, when the young shoots are nearly hard. The chief beauty of the plant resides in its graceful manner of growth and its light airy foliage. Its flowers are pallid and not dense enough to produce a handsome effect. When in fruit its large loose panicles of deep purple glaucous berries are ornamental enough; but their acid taste belies their tempting appearance. The species is unable to bear the winters of London, without the protection of a greenhouse.

310. Epidendrum quadratum. *Klotzsch, in Allgem. Gartenzeitung,* 1850, p. 402. An epiphyte from Central America, with racemes of brownish green flowers, and a dirty white lip dotted with red. Flowered with Mathieu and Allart of Berlin.

This seems to be very nearly the same as *E. varicosum* and *Lunæanum,* if not identical. But, according to the description, the leaves are much more narrow.

311. Pitcairnia fulgens. *Decaisne.* A stove herbaceous plant of the order of Bromeliads. Native of Guadaloupe. Flowers crimson.

Leaves spiny at the base, mealy beneath, as is the flower stem; raceme very close, with great pale green smooth bracts longer than the calyx; petals straight, two inches long, rich scarlet, linear-oblong, rounded, concave, with a crenated scale at the base. One of the Linden Collection seems to be handsome.

L. Constans, del. & Zinc.

Printed by C.F.Cheffins, London.

[PLATE 46.]

THE DARWIN BERBERRY.

(BERBERIS DARWINII.)

———◆———

A hardy Evergreen Shrub, from PATAGONIA, *belonging to the Natural Order of* BERBERIDS.

Specific Character.

THE DARWIN BERBERRY. Spines radiating, 5-parted, covered as well as the branches with a close rusty fur. Leaves simple, evergreen, wedge-shaped, 3-toothed, sometimes with another tooth or two at the side, spiny. Racemes dense, pendulous. Pistil flask-shaped, narrowed at the base, with a long style.

BERBERIS *DARWINII ;* spinis radiatis 5-partitis ramisque ferrugineo-tomentosis, foliis simplicibus sempervirentibus cuneatis 3-dentatis nunc dente uno alterove auctis spinosis, racemis densis pendulis, pistillo lagenæformi basi angustato stylo elongato.

Berberis Darwinii ; *Hooker, Icones plantarum,* t. 672 : *Lindley, in Journal of Hort. Soc.* vol. v., p. 6.

As far as our knowledge of this most beautiful bush extends, it must be regarded as the best hardy evergreen that has been imported for many years; scarcely inferior in horticultural value to a laurel or a holly. It is thus mentioned in an account of Evergreen Berberries cultivated in Great Britain :—

"Chiloe and Patagonia furnished this to Mr. T. Lobb, whose seeds have enabled Messrs. Veitch and Co. to raise it. Mr. Darwin also found it in Chiloe ; Bridges in Valdivia and Osorno.

"It forms an evergreen shrub three to five feet high, of extraordinary beauty, and conspicuous for its ferruginous shoots, by which it is at once recognised. The leaves are of the deepest green, shining as if polished, not more than three quarters of an inch long, pale green, with the principal veins conspicuous on the under side, with three large spiny teeth at the end, and about one (or two) more on each side near the middle. Although small, the leaves are placed so near together that the branches themselves are concealed. The flowers, which have not been yet formed in England, are in erect racemes, and of the same deep orange yellow as in the Box-leaved species.

" Mr. Veitch informs me that this plant appears to be decidedly hardy : as is probable, considering that it grows naturally near the summer limits of snow upon its native mountains.   It is now three feet high, and Mr. Lobb says it is, when a large plant, the finest he ever saw of the genus, in which I have no doubt that he is right."

To this we find nothing to add.   The coarse hairs that protect the spines and branches, but do not extend to the leaves, which are remarkably smooth and shining, distinguish the species at the first glance.

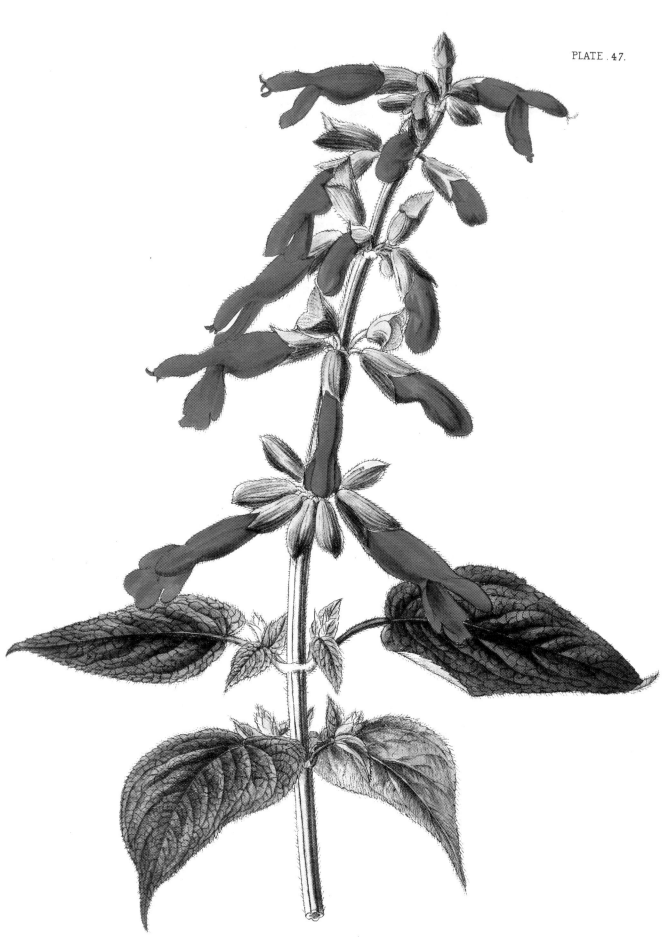

L. Constans. del & Zinc.

Printed by C.F.Cheffins, London.

[PLATE 47.]

THE GESNERA-FLOWERED SAGE.

(SALVIA GESNERÆFLORA.)

——◆——

A magnificent Greenhouse Herbaceous Plant, from CENTRAL AMERICA, *belonging to the Order of*
LABIATES.

Specific Character.

SECT.—Calosphace, longifloræ.—*Bentham.*

THE GESNERA-FLOWERED SAGE. The habit quite that of S. fulgens ; but the upper lip of the corolla flatter and less shaggy, the tube longer, the style less feathery, the flowers far more abundant and conspicuous.	SALVIA *GESNERÆFLORA ;* omninò facie S. fulgentis, corollæ galeâ planiore minus hirsutâ, tubo longiore, stylo minus plumoso, floribus copiosioribus magisque conspicuis.

Salvia Gesneræflora *of the Gardens.*

THERE is great difficulty in saying in what precise particulars this differs botanically from *S. fulgens,* or *Cardinalis;* the habit, foliage, and mode of flowering are the same in both, nor is there any difference in the flowers beyond what is pointed out in the foregoing character. Nevertheless the two plants are in a horticultural point of view quite distinct. This flowers all through the autumn and winter, *S. fulgens* is a summer species. The latter, handsome as it sometimes becomes, is no favourite on account of its incurably bad habit of becoming shabby and casting its blossoms. This on the other hand is of vigorous constitution, holds its flowers as well as a Gesnera, after which it is happily named, and has a fine rich brilliant green foliage.

The plant from which the accompanying figure was taken was struck from a cutting obtained at Syon, where it had been raised from Mr. Purdie's Colombian seeds. It formed a large bush more than three feet high, in a cold conservatory, and was a blaze of scarlet from November till April. During that time it was twice sent to London for exhibition, and on neither occasion exhibited the least appearance of having suffered in consequence.

It is struck readily from cuttings, and may be as common as a Stock or an Anemone, wherever a little greenhouse shelter and plenty of pot room can be given to it.

PLATE 48.

L. Constans, del. & Zinc.

Printed by C.F.Cheffins, London.

[PLATE 48.]

THE PALLID CATTLEYA.

(CATTLEYA PALLIDA.)

—◆—

A fine Stove Epiphyte, from the WEST OF MEXICO, *belonging to the Natural Order of* ORCHIDS.

Specific Character.

THE PALLID CATTLEYA. Pseudobulbs long, furrowed, one-leaved. Leaves wavy, oblong, blunt, emarginate. Flowers solitary, growing out of a very large spathe. Sepals lanceolate, petaloid.   Petals oblong, wavy, 4 times as broad.   Lip oblong, emarginate, rather wavy, hooded at the base for a little way.

CATTLEYA *PALLIDA ;* pseudobulbis elongatis sulcatis monophyllis, foliis undulatis oblongis obtusis emarginatis, floribus solitariis e spathâ maximâ enatis, sepalis lanceolatis petaloideis, petalis oblongis undulatis quater latioribus, labello oblongo emarginato subundulato basi cucullato lævi.

THIS is the Cattleya mentioned in Hartweg's Journal (*Journal of the Horticultural Society*, vol. i. 183) as having been found near Tepic, beyond which that miserable document gives no information.   It is nearly related to the Moss Cattleya, from which it differs in having very long furrowed pseudobulbs and flowers, without any indication of coloured veins.   The flowers are larger too than usual, the lip much less wavy, and the leaves weak and undulating instead of stiff and firm.

It is not so handsome as the generality of the species of this favourite genus ; but it is nevertheless a fine ornament to the orchid house.

H *

312. MAXILLARIA CONCAVA. *Lindley.* A pale yellow-flowered Epiphyte, from Guatemala, belonging to Orchids. Blossoms in November. (Fig. 159, diminished, with a magnified view of the lip.)

One of the less interesting of the racemose Maxillarias.

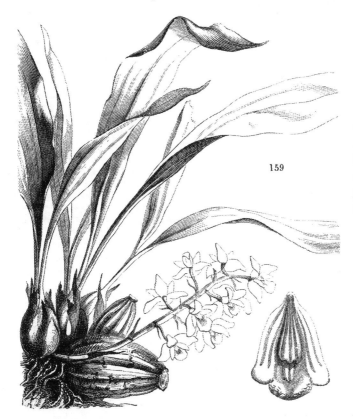

159

The flowers are pale yellow; the lip is almost truncate, concave, bluntly 3-toothed, with the middle lobe somewhat fleshy, and tuberculated at the edge; marked with rose-coloured veins, with a long narrow ridge in the middle, 3-lobed at the point. It is nearest *M. bracteata,* but its flowers are smaller, the bracts very small and bristly, and the lip of quite another form.

313. PERSEA GRATISSIMA. *Gærtner.* (*aliàs* Laurus Persea *Linnæus.*) A tree from the West Indies, where it produces the fruit called the Alligator Pear. Flowers green, downy, in panicles. Belongs to the order of Laurels.

"The 'Avocado,' or 'Alligator Pear,' yields a fruit never, that I am aware, known to be produced in Europe; nor am I aware that it has ever flowered in our stoves, save at Syon and Kew. In the West Indies it is highly valued, and cultivated, and in tropical America generally. It is presumed to be an aboriginal of these countries; though some say imported to the islands from the South American continent. Why called *Alligator Pear* is not very evident. Perhaps the first word is a corruption of *Aguacate,* one of the names by which, according to Ulloa, it is known in Lima. The fruit is pear-shaped, yellow or brownish-green, often tinged with deep purple. Between the skin and the hard seed is a pale butyraceous substance, interspersed with greenish veins, and this is much eaten by all classes of people; its taste somewhat resembling butter or marrow, and hence is called the 'vegetable marrow:' and this is so rich and mild that most people make use of some spice or pungent substance to give it poignancy: and wine, sugar, lime-juice, but mostly pepper and salt, are used. However excellent when ripe, the *Avocado* is very dangerous if pulled and eaten before maturity; being known to produce fever and dysentery. 'If you take the stone of the seed,' says Barham, 'and write upon a white wall, the letters will turn as red as blood, and never go out till the wall is white-washed again, and then with difficulty."— *Bot. Mag.*, t. 4580. There are some wax models of this fruit in the Crystal Palace, among the articles from the British West Indian Colonies.

160

161

314. BROWALLIA JAMIESONI. *Bentham.* A handsome greenhouse shrub, with orange-yellow flowers, belonging to Linariads. Native of Peru. Blossoms in the autumn. Introduced by Messrs. Veitch & Co. (Fig. 160.)

This very pretty shrub inhabits various parts of the kingdom of New Grenada, near Loxa, &c. It forms a neat dark-leaved bush, with stalked, oblong, wrinkled leaves, and solitary, axillary, orange-yellow flowers, which, when the plant is very healthy, are collected into small terminal corymbs. The species was introduced by Messrs. Veitch, who alone have hitherto produced specimens at our exhibitions.

315. ACACIA VERTICILLATA. *Willdenow.* A loose, straggling, prickly, greenhouse shrub. Native of Van Diemen's Land. Flowers, light yellow, in March and April. (Fig. 161.)

How this differs from *A. Oxycedrus* will be obvious upon comparing the present cut, with that at p. 43 of our last number. It is found in the same country, abounding all over the island, where it assumes many forms ; sometimes having broad leaves, and a stout almost erect habit, sometimes having almost awl-shaped leaves, and not possessing stiffness enough to support itself. It bears long, narrow, blunt, curved pods. *A. Riceana*, figured at p. 42, approaches it more nearly than any thing else ; but that has a much more graceful habit, and its paler flowers are so disposed that each may be seen separately upon the long drooping spikes, while here, on the contrary, even when old as in our figure, they always have a comparatively compact arrangement ; and when young, are collected into close oblong spikes. Although treated as a greenhouse plant, this species ought to be hardy in the S Western counties.

316. EUONYMUS FIMBRIATUS. *Wallich.* A beautiful evergreen greenhouse bush, from the Himalaya. Flowers, green. Belongs to the order of Spindle trees (*Celastraceæ*). (Fig. 162.)

Although, in this country, fine flowers are more looked for than beautiful foliage, yet in the present case the very handsome appearance of the plant, now for the first time figured, ought to satisfy even English fastidiousness. We can scarcely do better than transcribe Dr. Wallich's description of it :—

"Specimens of this beautiful species were communicated from the Servalik mountains, by Dr. Govan ; and from Shreenugur, by Kamroop. Probably a tree. *Branches* round, slender, gray ; while young alternately compressed ; all parts smooth. *Buds* axillary and terminal, oval, acute, consisting of ovate, obtuse, imbricating scales. *Leaves* opposite, ovate, terminated by a lanceolar, long acumen, margins most elegantly marked with narrow, linear, lanceolate, sharp, slightly incumbent, parallel, and approximate serratures, which are two or three lines long, and sharply denticulate, or serrulate ; base rounded, or acute, nearly entire ; the lower surface with a strong rib and oblique nerves, from three to five inches long. *Petiole* half-an-inch long, and furrowed. *Peduncles* lateral, approximate on the young shoots, filiform, a little flattened, two or three inches long, divided into five or six long slender rays, each bearing a simple or compound fascicle of tetrandrous *flowers*. *Calycine* segments oblong, obtuse. *Stamina* very short. *Capsule* large, turbinate, depressed, furnished with from two to five lanceolate, tapering, vertical, horizontally spreading wings, which are sometimes two-thirds of an inch long, and as broad at the base as the capsule itself is deep.

" *Obs.*—It is impossible to confound this with any other species. While young the leaves are lanceolate, less deeply, but distinctly duplicato-serrate ; when old they become broad ovate, elegantly fringed with numerous narrow, deep, dentate serratures. The capsules are large and leathery."

In the winter the large leathery seed vessels open and display the rich orange-coloured seeds, which themselves produce a sufficiently gay appearance.

317. HELLEBORUS ATRO-RUBENS. *Waldstein and Kit-aibel.* A hardy herbaceous plant, with dull purple flowers, appearing early in the spring. Native of Hungary. Belongs to the Crowfoots.

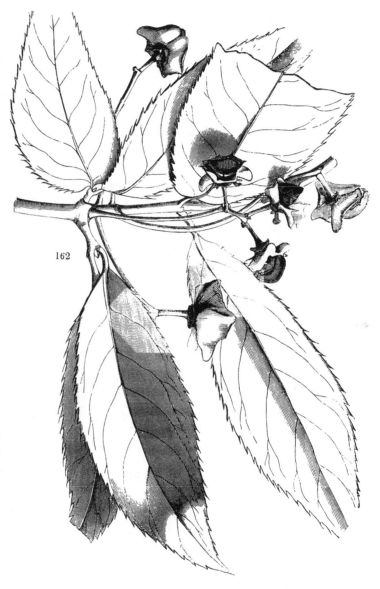

162

" A really handsome and hardy herbaceous flowering plant, blossoming when flowers are more especially welcome visitors, in February and March. The blossoms are large, spreading, at first rather a dark purple (hardly dark enough to justify the name *atro-rubens*), gradually changing to green as the fruit advances to maturity. It inhabits woods and bushy places in the mountain districts of Croatia, and is especially abundant about Korenicz. Root a branched tuber or cormus, throwing down very numerous long fibres. Stem erect, herbaceous, dichotomously branched, glabrous, obsoletely angular. Root-leaves coming to perfection after the flowers, pedate, shining, the lobes lanceolate, reticulated, finely serrated, shining, paler beneath. Stem-leaves with a sheathing base, almost sessile, less divided : uppermost ones or bracteas at length lanceolate, undivided. Peduncles mostly terminal and two-flowered. Sepals broad ovate, almost rotundate, spreading, dull but rather dark red-purple, persistent and changing to dull pale brownish-green. Petals wedge-shaped, a short compressed tube, open at the mouth. Stamens numerous, yellow. Pistils five. Ovaries tapering into styles as long as the stamens. Stigma clavate, hairy."—*Bot. Mag.*, t. 4581.

318. Skimmia Japonica. *Thunberg.* (*aliàs* Limonia Laureola *Wallich.*) A half hardy, fragrant, evergreen shrub, belonging to Citronworts. Found in China and Japan, and on the Himalayan Mountains. Flowers, pale green. (Fig. 163.)

This interesting evergreen bush has now established itself in Cornwall, and is procurable in the nursery of Messrs. Standish and Noble. We cannot see that the Himalayan and Chinese forms differ in any respect; for although Zuccarini says the flowers of the Chinese plant are white, edged with rose, yet the Chinese specimens furnished to us are green, exactly as represented in Dr. Wallich's *Plantæ Asiaticæ rariores.* We have therefore only to add to our assurance that it is a bush with deliciously fragrant blossoms, the following extract from Zuccarini's *Plantæ Japonicæ* :—

" The *Skimmi* is an evergreen shrub, found throughout Japan among mountains, and shaded by forests, but the plants are nowhere numerous, and always scattered, which renders them somewhat rare. We found it near Nangasaki, on the mountain Kawara, 594 yards above the level of the sea. Kaempfer was wrong in calling it a large tree, for in a wild state it scarcely grows above three or four feet high, and its branches generally incline towards the earth. In a cultivated state it becomes taller, the branches crowded, and almost in whorls of three or four together, growing stiffly from the stem. The leaves last three or four years, and each year is known by the distance of the tufts. The wild plants are distinguished by sharper leaves, which are closely spotted with transparent glands, like those of the orange tree, or the Tutsan. The flowers appear, in terminal panicles, from the beginning of March to the end of April. They are white, and of a reddish colour at the edge of the under side of the petals. The perfume is very agreeable, not unlike that of *Daphne odora*, and is particularly strong in the evening. The round, bright red berries (white in a cultivated variety) resemble those of the hawthorn, and contain four cartilaginous stones. The fruit ripens in October, and does not fall till towards the end of December. The leaves have an aromatic and acrid flavour.

163

" The *Skimmi* is generally cultivated in Japan in gardens, and around temples. Its evergreen bright leaves, its numerous and graceful bunches of flowers, displayed from the tops of the branches from the beginning of spring, its fragrance, and at the end of autumn its beautiful red berries, entitle it to a high rank among ornamental plants. It is increased by cuttings or layers. Although in our climate (Holland) it is difficult to preserve in the open air, it thrives perfectly in the green-house, where, along with Camellias, it enhances by its perfume the beauty of those scentless shrubs. However, the Japanese and the Chinese reckon it among venomous plants; and the name *Sikimi* signifies also mischievous fruit."

319. Acacia cyanophylla. *Lindley*. A noble Swan River tree, with long glaucous leaves, and spikes of bright yellow flower-heads. Blossoms in February and March. (Fig. 164.)

According to Preiss this plant inhabits wet sandy flats, near Swan River, where it is called Black Wattle, and forms a small straggling tree from 12 to 18 feet high: he adds, that the leaves of the wild plant are much smaller and narrower than in the cultivated. In this country it is one of the finest decorations of the conservatory where room can be afforded. Its drooping branches are loaded with long glaucous, almost blue, leaves,—themselves handsome objects ; and in the first months of the year, it pours forth in profusion its long spikes of deep yellow round flower-heads. Some of the leaves are above a foot long. The pods are said to be very narrow, from three to five inches long, and contracted between the seeds ; we have not seen them.

320. MORMODES BARBATUM. (*aliàs* Mormodes atro-purpurea *Hooker*.) An Orchidaceous epiphyte, from Central America. Flowers dark purple. Introduced by Warczewitz. Blossomed by J. D. Llewelyn, Esq.

This plant, published lately in the *Botanical Magazine*, t. 4577, under the name of *Mormodes atro-purpurea*, is quite different from the plant so called in the *Botanical Register*, t. 1861. It differs not only in the longer and looser spike, but in the form of the lip, which is not at all 3-lobed, or rather 5-lobed, but quite entire, and moreover covered with long hairs instead of being smooth. Sir Wm. Hooker gives the following definition and account of it :—

M. pseudo-bulbis oblongis squamis amplis imbricatis pallidis fusco-marginatis vaginatis, foliis . . ., floribus pendulis unicoloribus, sepalis petalisque arcte reflexis ovato-lanceolatis marginibus

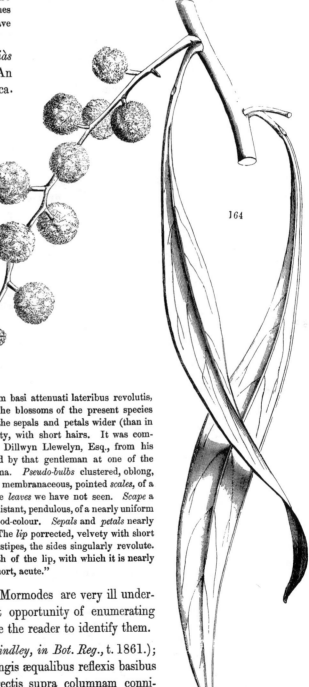

164

revolutis, labelli late obcordati velutini in stipitem basi attenuati lateribus revolutis, columna oblique torta breviter acuminata. " The blossoms of the present species are of a uniform dark purple or blood colour, the sepals and petals wider (than in *M. lentiginosa*), the lip much broader and velvety, with short hairs. It was communicated in January, 1851, by our friend J. Dillwyn Llewelyn, Esq., from his collection at Penllergare, having been purchased by that gentleman at one of the sales of plants of Mr. Warczewitz, from Panama. *Pseudo-bulbs* clustered, oblong, striated, the old ones entirely sheathed by large, membranaceous, pointed *scales*, of a pale straw-colour, edged with dark brown. The *leaves* we have not seen. *Scape* a foot high, rounded, articulated. *Flowers* rather distant, pendulous, of a nearly uniform dark purple-brown, or between chocolate and blood-colour. *Sepals* and *petals* nearly uniform, ovato-lanceolate, their sides reflexed. The *lip* porrected, velvety with short hairs, broadly obcordate, tapering below into a stipes, the sides singularly revolute. *Column* pale, purplish-brown, not half the length of the lip, with which it is nearly parallel, but it has an oblique twist ; the apex short, acute."

As the species of the curious genus Mormodes are very ill understood, we avail ourselves of the present opportunity of enumerating them, with such short notes as will enable the reader to identify them.

321. MORMODES ATROPURPUREUM (*Lindley, in Bot. Reg.*, t. 1861.); racemo oblongo denso, sepalis lineari-oblongis æqualibus reflexis basibus lateralium paulo obliquis, petalis ovatis erectis supra columnam conni-

ventibus, labello replicato retrorsum arcuato cuneato trilobo : lobis lateralibus deflexis venosis inter-medio carnosiore cuspidato subtrilobo.—*Central America.*—Flowers deep purple, on a close erect raceme. Lip quite smooth, 3-lobed, veiny at the edge, the middle lobe slightly 3-fid, fleshy, with the divisions rounded, that in the middle being longer and cuspidate.

322. MORMODES LINEATUM (*Bateman, in Bot. Reg.* 1841, *misc. no.* 107. 1842, *t.* 43.) ; racemo elongato multifloro, sepalis petalisque oblongo-linearibus abruptè acutis margine reflexis, labello lineari incurvo carnoso sparsim piloso versus basin utrinque dente nunc brevi nunc elongato aucto, columnæ dorso et margine pubescente.—*Guatemala.*—The flowers are deliciously fragrant; when they first appear they are dull olive green ; they afterwards acquire a bright warm yellow tint, and the markings upon them increase in intensity till they have become orange-red.

323. MORMODES CARTONI (*Hooker, in Bot. Mag., t.* 4214.) ; pseudo-bulbis elongatis teretibus articulatis vaginatis apice di-triphyllis, foliis lineari-lanceolatis acuminatis, spica elongata multiflora, sepalis petalisque patentibus conformibus oblongo-lanceolatis acutis, labello oblongo torquato basi angustato infra medium utrinque unidentato marginibus reflexis apice aristato-acutis, columna antheraque cuspidato-acuminatis.—*Central America.*—From the collection sent home by Mr. Purdie, from the interior of Santa Martha, at the foot of Sierra Nevada. It first flowered at Syon Gardens, and it is nearest to *M. aromaticum,* but it is at once distinguishable by the lip and various other discrepancies.

    Pseudo-bulbs almost a span long, clustered, subcylindrical, articulated, and sheathed at the joints by the membranaceous bases of the old leaves. Perfect leaves three to four, are produced from the apex of the bulb, a foot or more long, narrow, linear-lanceolate, membranaceous, striated, acuminated. Scapes one or two from an articulation of the pseudo-bulb, erect, bearing a rather oblong spike of numerous rather gay-coloured flowers. The sepals and petals are nearly uniform in size and shape, much spreading, almost reflexed, oblong-lanceolate, acute, yellow with red longitudinal streaks. Lip equal in length with the petals, but singularly obliquely twisted, of a pale yellow colour, with a few red interrupted streaks ; the form is an irregular oblong, tapering at the base, with a short blunt tooth on each side below the middle, the apex very acute, almost aristate. Column slightly oblique, tapering at the extremity into a long subulate point. Anther-case corresponding with it, and applied to the anterior face. Pollen-masses two, each with an obscure fissure, attached to a broad, curved, strap-shaped appendage, and that again by its base to a large gland. —*Hooker.*

    Flowers extremely variable. Found on Erythrinas, in the temperate regions of the Sierra Nevada of Sta. Martha. The sepals are flesh-coloured ; lip fleshy, rose. The same plant was found by Mr. Funck, at Minca, near Sta. Martha, in the province of Rio Hacha, at the height of 4000 feet.

324. MORMODES LENTIGINOSUM (*Hooker, in Bot. Mag, t.* 4455.) ; pseudo-bulbis oblongis, foliis lanceolatis striatis, scapo radicali racemoso, floribus remotis pendulis punctis rufo-fuscis irroratis, sepalis arcte reflexis oblongis acutis marginibus revolutis, petalis conformibus, labelli obovati lateribus evolutis, columna oblique torta apice acuminata.—*Central America.*

    The young pseudo-bulbs are almost globose, leafy ; the old ones are oblong, terete, and partially sheathed with the withered membranous bases of the old leaves. Leaves, in the plant before us, a span long, lanceolate, membranaceous, striated. Scape from the base of a pseudo-bulb, a foot or a foot and a half high, racemose, flexuose where the remote drooping flowers are inserted. Bracteas small, membranaceous, acuminated. The general colour of the flower is pale reddish-brown, everywhere sprinkled with dark-coloured dots. Sepals and petals acute, with margins singularly recurved, the former (sepals) refracted, the petals having an opposite direction, the same as that of the column and lip. Lip rather large, fleshy, obovate, entire, the sides entirely curved back, revolute, almost like the flaps of a saddle. Column shorter than the lip, semiterete, curved, and having a singularly oblique twist, acuminated. Anthers taking the same shape as the apex of the column, and hence much acuminated too ; the colour is a deeper red than the rest of the flower.—*Hooker.*

325. MORMODES BUCCINATOR (*Lindley, in Bot. Reg.,* 1840, *misc.* 9.) ; racemo stricto oblongo, sepalis lineari-oblongis lateralibus reflexis dorsali petalisque ovali-lanceolatis erectis, labello ungui-culato carnoso nudo subrotundo cuneato apiculato utrinque emarginato lateribus in buccinæ formam

revolutis.—*Venezuela*.—Flowers pale-green, with an ivory-white lip, whose sides are so rolled back as to give it the appearance of a trumpet. The column is twisted, sometimes to the right, sometimes to the left. The habit and general appearance of the flowers, except in colour, is that of *M. atrópurpureum*.

When this was first published nothing was known of its native country. At a later period it was said to be Mexican, upon some bad garden authority. We find it, however, with brown flowers, among Schlim's Venezuela plants, No. 67, gathered at San Cristobal at the height of 2500 feet above the sea.

326. MORMODES AROMATICUM (*Lindley, in Bot. Reg.* 1841, *misc.* 162, 1843, t. 56); racemo brevi erecto, sepalis petalisque subrotundo-ovatis acutis secundis concavis, labello angustè cuneato convexo laciniâ intermediâ triangulari acuminatâ cucullatâ.—*Mexico*.—This has flowers with a pale pinkish ground sown thickly with dull wine-red specks, and a powerful odour like that of aromatic vinegar. It differs from *M. pardinum* in the small size of its flowers, and the very dissimilar form of the lip and floral envelopes.

327. MORMODES PARDINUM (*Bateman, Orch. Mexic.*, t. 14; *Hooker, Bot. Mag.*, t. 3900; aliàs *Cyclosia maculata* Klotzsch, in Gartenzeit., no. 39, 1838.—Var. *Unicolor* Hooker, l. c. t. 3879; *Catasetum citrinum* Hort.); foliis elongatis lanceolatis, racemo ascendente elongato multifloro, sepalis petalisque lanceolatis acuminatis subsecundis, labello plano conformi acutè tridentato unguiculato.—*Mexico*.—A beautiful species, with bright yellow flowers, spotted with rich brown in one variety; whole coloured with no spots whatever in the other.

328. MORMODES LUXATUM (*Lindley, in Bot. Reg.* 1842, *misc.* 66, 1843, t. 33.); foliis longissimis angustis subtùs glaucis racemo oblongo pluriès longioribus, sepalis ovato-lanceolatis petalisque oblongis concavis margine subscariosis carnosis incurvis, labello hemispherico concavo obsoletè trilobo apiculato supra columnam cucullato, caudiculâ apice tuberculatâ. — *Mexico*. —Flowers large, as much as 3½ inches in diameter, pale lemon-colour, powerfully aromatic, with somewhat the appearance of an Anguloa. The manner in which the customary arrangement of the parts of fructification is disturbed is very curious. We have not seen the plant in collections for a long time.

165

329. ACHIMENES VISCIDA. (*aliàs* Cheirisanthera atrosanguinea *of the Gardens*.) A hothouse herbaceous plant with

viscid woolly leaves, and red and white flowers. Native of ———— ? Flowers nearly all the year round. (Fig. 165.)

A. viscida; undique pilis viscidis tomentosa, foliis ovatis oblongisque crenatis, cymis pedunculatis axillaribus pauci-floris, corollâ basi supernè gibbosâ tubo rectiusculo limbo 5-lobo laciniis rotundatis subæqualibus, ovario hirsuto.

In what work the name which this plant bears in the gardens has been proposed we have failed to discover. It was imported from the continent, and is believed to be one of Linden's plants, but can hardly be the *Achimenes atrosanguinea* of Morren. A half inferior ovary, a complete narrow annulus, a fifth abortive stamen in addition to the four perfect ones, and the nearly equal limb of the corolla, seem to pronounce the plant an Achimenes, although the habit is more that of an Isolome. It is a soft and not handsome hothouse plant, from two to three feet high, closely covered with long, slender, delicate hairs, from whose points a green viscid substance is continually exuding. The calyx is regularly five-lobed; the corolla of a uniform deep crimson, with the inside of the tube, and the orifice of the throat, nearly white; at the base of the tube is a circular swelling which is more considerable on the upper than the under side, and upon the inner face of this swelling stand five stamens with broad fleshy bases, the fifth of which is generally straight and sterile, but sometimes as perfect as the others. The stigma is slightly two-lobed—the lobes expanding right and left as in other species of Achimenes.

330. HILLIA PARASITICA. *Jacquin.* (*aliàs* H. longiflora *Swartz.*) A handsome hothouse creeper, with long trumpet-shaped, cream-coloured, flowers. Belongs to Cinchonads. Native of the West Indies. (Fig. 166.)

Jacquin, who first discovered this plant overrunning trees and old walls in the dense damp woods of Mount Calebasse, in Martinique, called it a Parasite. It however deserves that name no more than ivy, to which it may be compared as to its habit; striking roots into soil, or clinging to rotten bark when it comes in contact with it, or rising feebly from the ground if there is nothing to cling to. It is very rarely seen in gardens; and yet it is one of the easiest of plants to cultivate, requiring the same treatment as would suit the now common *Stephanote.* Its leaves are firm, rather fleshy, deep green, and handsome. The flowers are four inches long, with a slender tube,

166

and six reflexed divisions; towards the mouth the tube of the corolla becomes inflated like the mouth of a trumpet; they are a delicate cream-colour when first opened, but soon acquire the peculiar yellow tint observed in *Gardenia Fortuni*, and other species of that genus. According to De Candolle, this is found not only in Martinique, but in Guadaloupe, Jamaica, Cuba, and the hot parts of Mexico.

Printed by C.F.Cheffins.London.

L. Constans. del. & Zinc.

[PLATE 49.]

THE BOX-LEAVED CANTUA.

(CANTUA BUXIFOLIA.)

———◆———

A beautiful half-hardy Shrub, from PERU, *belonging to the Natural Order of* POLEMONIADS.

═══════════════

Specific Character.

THE BOX-LEAVED CANTUA. Leaves oval, acute, smooth or downy, hardened at the base ; sometimes three-lobed or otherwise lobed. Panicles loose, downy, corymbose. Calyx downy, blunt at the base, more than thrice as short as the corolla. Corolla a long tube with a concave limb and obcordate segments. Style projecting.

CANTUA *BUXIFOLIA ;* foliis ovalibus acutis glabris v. pubescentibus basi induratis nunc trilobis, paniculis laxis corymbosis tomentosis, calyce pubescente basi obtuso corolla plus triplo breviore, corollâ longè tubulosâ limbo concavo laciniis obcordatis, stylo exserto.

─────────────

Cantua buxifolia : *Lamarck Dict.,* 1. 603. *Bentham in De Cand. Prodr.,* 9. 321. *Bot. Mag.,* t. 4582 ; *aliàs* Periphragmos dependens *Ruiz and Pavon, Fl. Chilensis,* II. 18, t. 133.

═══════════════

SINCE the introduction of the Fuchsia and the China Rose, our gardens have received nothing so remarkable as this plant. Long since made known to botanists, and sought for by every collector that has visited the temperate parts of South America, it has at last been obtained by Mr. W. Lobb, from the mountains of Peru, for Messrs. Veitch of Exeter, who have flowered what is now represented. The blossoms appear in profusion in the month of May, and are fully four inches long, with a crimson and yellow tube, vivid sanguine in the bud, and rich rose colour when expanded, with a lighter tint in the inside.

There is no reason to doubt that this will be as hardy, and as cultivable, as the Fuchsia itself, and we may expect to see it in a few years in every cottage garden.

In his enumeration of Polemoniads, in De Candolle's Prodromus, Mr. Bentham has reduced to this species the *Cantua tomentosa* of Cavanilles ; and Sir W. Hooker has gone farther, in the "Botanical Magazine," by adding as synonymes the *Cantua ovata* of Cavanilles, and *uniflora* of Persoon. We however believe that these are so many distinct species.

It is doubtless true that *Cantua buxifolia* is a variable plant, more or less downy, and having flowers either crimson and yellow as this is, or white and yellow, or perhaps merely yellow. All these forms may be expected to appear from the same batch of seeds. In fact, among Mr. W. Lobb's dried specimens, no fewer than six different numbers are occupied by the forms of the same species, this *C. buxifolia*. But the materials before us lead to the inference that other forms of the genus exist in temperate South America, which are specifically distinct from *C. buxifolia*, and from each other.

In the first place we have a Peruvian plant collected by Dombey, and distributed by the Paris Herbarium, under the name of " *C.* grandiflora, No. 382." This, which is nearly entirely glabrous, has much shorter flowers, and blunter leaves than *C. buxifolia*, the calyx being almost half as long as the corolla tube ; it is probably *C. ovata*.

Among Bridges' last Bolivian Collection, is a shrub with leaves and calyxes covered all over with a viscid glandular pubescence, an extremely narrow crimson streaked corolla and calyx, the latter tapering to the base, and an inconsiderable limb, shorter than the projecting stamens. That we presume to be *C. tomentosa*.

Finally we have in the same collection, a species with pubescent leaves, with a great tendency to become round at the point, flowers growing singly at the end of short lateral branches, and glabrous calyxes, almost half as long as the tube of the yellow corolla ; and to this the name of *C. uniflora* seems to belong.

These plants are not unlikely to appear in our gardens now that the importation of seeds has commenced, and at all events should be diligently sought for by those who have correspondents in either Bolivia or Peru.

PLATE 50.

L. Constans del.& Zinc.

Printed by C.F. Cheffins London.

[PLATE 50.]

THE CRIMSON WATER-LILY.

(NYMPHÆA RUBRA.)

———•———

A Stove Aquatic, from the E. INDIES, *belonging to the Natural Order of* WATER-LILIES.

═══════════════

Specific Character.

THE CRIMSON WATER-LILY. Leaves roundish, ovate, slightly peltate, toothed, deeply split at the base, downy on the underside. Flowers crimson. Sepals 7-ribbed. Stigmas 15.	NYMPHÆA *RUBRA;* foliis subpeltatis ovato-subrotundis dentatis basi fissis subtùs pubescentibus, floribus sanguineis, sepalis 7-nerviis, stigmatibus 15.

———

Nymphæa rubra: *Roxburgh's Flora Indica,* ii. 576; *aliàs* Castalia magnifica: *R. A. Salisbury, Paradisus Londinensis,* t. 14.

═══════════════

THIS brilliant aquatic, though an old inhabitant of our gardens, is still a rarity, appearing only in first class collections. Nor has it been fortunate in the artists who have attempted to fix its likeness on paper; the early figure in the "Botanist's Repository" is particularly unsatisfactory. We have, therefore, gladly availed ourselves of the opportunity of producing a true representation of a fine specimen which flowered in March last at Syon.

Roxburgh merely says of this plant that it is a native of India, flowering during the rainy season, and by no means so common as the *Nymphæa Lotus,* nor do we find other details in the works of Indian writers.

It is probable that more species than one may be included in this name, for Roxburgh mentions a small rose-coloured variety with from twenty to twenty-five stamens, and Dr. Wight figures as *N. rubra,* a plant with at least sixty long narrow yellow stamens. Neither of these corresponds with that before us, which we presume to be the common Indian plant; in which we find, strongly seven-ribbed sepals, crimson inside, succeeded by broad satiny spreading crimson petals, the central of which are linear, blunt, erect, curved inwards, and gradually passing into the crimson stamens, which they nearly conceal. The stigmatic apparatus consists of fifteen papillose rays, which are free and smooth at their extremities, curved inwards and fleshy, surrounding a central nipple; as in *Nymphæa alba* and others. To these extremities it is desirable that anatomists should direct their attention, inasmuch as their peculiar construction indicates some very peculiar function. In *Nymphæa alba* they are deep yellow, firm like wax, with a strong even epiderm, and are filled with a soft loose cellular substance, containing an abundance of large coarse scabrous hairs, sometimes half circular, sometimes straight, all placed parallel with the external surface. They are evidently analogous to the scabrous hairs so abundant in the air cells of Nymphæa. The yellow ends of the stigmatic rays of Nuphar do not contain this tissue. Nor is there anything in Victoria, much as that plant abounds in stellate internal hairs, which is identical with the extremities of the stigmatic apparatus of Nymphæa.

Other peculiarities are observable among water-lilies, and are, it must be supposed, connected with their vital functions, although we know not in what way. The pollen, for instance, varies greatly in some of the genera. In *Nymphæa rubra* it is simple, globose, and perfectly smooth; in *Nymphæa alba* it is similar, but the surface is slightly rough. In the yellow water-lily (*Nuphar lutea*), on the contrary, it is covered with such long points that the pollen-grains hold together in masses, like burs. On the other hand in *Victoria*, where the pollen is much larger, the grains are perfectly smooth and constantly grow together in threes or fours.

In the Crimson Water-lily the leaves are closely covered on the underside with a soft felt of delicate hairs, which are quite perceptible to the touch. Examined with the microscope the hairs are found to be simple attenuated smooth cones, with no tendency to branch or become stellate. Not a trace is perceptible on the leaf of those curious perforations in *Victoria* which have been mistaken for stomates, but which in reality are passages through the thickness of the leaf, and are altogether, as far as we know, *sui generis*. We may as well take the present opportunity of saying of these perforations that instead of being stomates, which are also present in *Victoria*, they are formed by a depression of two corresponding points of the upper and under surface of the leaf, and are at first closed by a transverse membrane. After a short time this membrane disappears, and a clear passage through the leaf is thus effected. Possibly this contrivance may be intended to allow the air to escape upwards, that would otherwise collect below the under surface of the leaves in *Victoria* in the spaces included by its deep ribs, and thus prevent that contact of water which may be assumed to be necessary to the health of that extraordinary aquatic.

PLATE 51.

Printed by CF Cheffins, London.

1. Constans, del. & Zinc.

[PLATE 51.]

THE HUMBLE PLEIONE

(PLEIONE HUMILIS.

———◆———

An Alpine Herbaceous plant, from NORTHERN INDIA, *belonging to the Natural Order of* ORCHIDS.

Specific Character.

THE HUMBLE PLEIONE. Pseudobulbs flask-shaped furrowed. Bracts oblong-lanceolate, petaloid, longer than the ovary, but afterwards shrivelling and drawn back, leaving the peduncle naked. Sepals and petals linear-lanceolate, spreading, flat. Lip hooded, emarginate, fringed at the edge, and with 6 distant fringed veins between which as many naked veins are interposed.

PLEIONE *HUMILIS;* pseudobulbis lagenæformibus sulcatis, bracteâ oblongo-lanceolatâ petaloideâ ovario longiore demum retracto pedunculo denudato, sepalis petalisque lineari-lanceolatis patentibus planis, labello cucullato emarginato necnon lineis 6 fimbriatis distantibus venis totidem coloratis interjectis.

Pleione humilis : *D. Don, Prodromus Floræ nepalensis,* p. 37 ; *aliàs* Epidendrum humile : *Smith Exot. Bot.* t. 98 ; *aliàs* Cymbidium humile : *Smith in Rees' Cyclopædia ; aliàs* Cœlogyne humilis : *Lindley, Collectanea Botanica,* p. 37.

THIS beautiful gem was originally found by Dr. Buchanan Hamilton in Upper Nepal, among moss, on the trunks of trees. Mr. Griffith met with it on the Bootan mountains in similar situations, in dense forests towards Santagong at the elevation of 8000 feet above the sea (Itinerary Notes, p. 158). Recently it has been sent from the Khasijah hills to Messrs. Veitch, by Mr. Thomas Lobb, who found it at a place called "Sanahda," at the height of 7000 feet.

It differs from the two species figured at our Plate 39, in the form of its pseudobulbs, in the narrowness of the sepals and petals, and especially in the long fringes that border the lip, and which also occur upon six of the veins on the inside of the lip. The bract too is quite different, petaloid and pale violet at first, then shrinking and shrivelling till it leaves the peduncle naked as in our figure, remaining at the base of the peduncle like an old-fashioned leather buskin.

The habit of these plants is so peculiar that it seems desirable to separate them from Cœlogyne, if any character can be found; and we think the membranous bracts, and strongly saccate lip with fringed veins of Pleione may be taken to offer a sufficient distinction from Cœlogyne with its horny or cartilaginous deciduous bracts, and lip merely concave at the base, with two or three continuous crests rising up from the veins. Perhaps too a more careful comparison than we have been able to make of the pollen of the two genera may furnish peculiarities of a more important kind.

At all events Pleiones constitute a group which can never be intermingled with the species of Cœlogyne proper. The following is an enumeration of the species now known :—

PLEIONE *D. Don.*

1. Pleione *maculata.* Plate 39 of this work.
2. ——— *lagenaria.* „ „
3. ——— *Wallichiana* (*aliàs* Cœlogyne Wallichiana *Lindley.*)
4. ——— *præcox* D. Don. A fine species with large purple flowers and a fringed lip.
5. ——— *humilis* of this plate.
6. ——— *diphylla ;* pseudobulbis oblongis medio constrictis, foliis geminis subcoriaceis acuminatis, pedunculo flore ter longiore, bracteâ obtusâ inflatâ apiculatâ ovarii longitudine, labello obtusè trilobo emarginato, venis fimbriatis 5-7 interruptis alterâ brevi adjectâ utrinque juxta apicem.

We have specimens of this plant from Mr. Griffith, who found it on the Khasyah Mountains, in shaded rocky places at Churra : and whose memorandum appears in his Itinerary Notes, p. 44, No. 684. From this it appears that the leaves are somewhat coriaceous, and grow in pairs on the summit of oblong pseudobulbs, contracted in the middle, and spotted with purple on a green ground. The flowers are said to be very handsome, and white ; the lip being stained and lined inside with violet and crimson, and decorated with from seven to nine lines of yellow fringes.

GLEANINGS AND ORIGINAL MEMORANDA.

331. Clematis graveolens. *Lindley.* A hardy climber from the north of India. Leaves pale green, deciduous. Flowers greenish yellow, heavy-scented. Flowers in the middle of summer. (Fig. 167.)

This pretty little plant was raised in the garden of the Horticultural Society, from seeds collected by Captain William Munro, in Chinese Tartary, and the Snowy Passes, at an elevation of 12,000 feet. In cultivation it proves to be a small slender climbing species, perfectly destitute of hairiness, except on the calyx and fruit. The leaves have very small ovate, three-lobed, leaflets, and long straggling footstalks. The flowers are solitary, at or near the extremity of the branches, pale yellow, rather pretty, but emitting a heavy smell, which, in a greenhouse, is more disagreeable than pleasant, but is not observed in the open air. It proves to be perfectly hardy in the severest winters; grows freely in any good loamy soil, and is easily increased by cuttings. The seed was sown in the garden of the Horticultural Society on the 17th of May, 1845; and the plant was in flower by the end of July, 1846. See *Journ. of Hort. Soc.*, vol. i., p. 307.

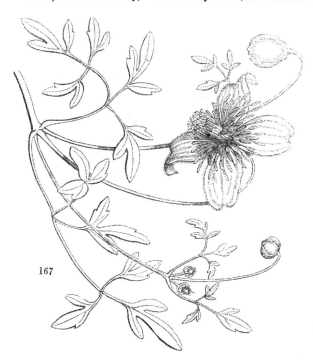

167

332. Bifrenaria Hadwenii. (*aliàs* Scuticaria Hadwenii *Hort.*) An Orchidaceous epiphyte, with pale buff flowers, mottled with brown on the sepals, and rose colour on the lip. Native of Brazil. Flowers in May. Introduced by Thomas Brocklehurst, Esq.

B. *Hadwenii;* floribus solitariis, sepalis petalisque undulatis acuminatis, labello cucullato emarginato subrepando intus pubescente cristâ 3-dentatâ.

We have only seen a solitary flower of this plant, which bears in some gardens in the North of England the name of *Scuticaria Hadwenii.* According to Mr. Wm. Pass, of Macclesfield, from whom the specimen came, " Mr. Hadwen was the first to receive the plant from Mr. de Becca, of Rio de Janeiro, who has since sent it to Mr. Brocklehurst. The habit of the plant is very distinct from *Scuticaria Steelii*, having leaves or stems not more than twelve to fourteen inches long, which gives it the appearance of a Brassavola. From what I learn, the flowers are solitary and on long stems like the one sent." The species differs in little from Bifrenaria, and is probably allied to *B. inodora*. The flowers when expanded are about three inches in diameter. The sepals are convex, those at the side being only slightly extended into a chin. The petals have a similar form, are rather inclined to turn back at the edges, and converge over the column, which is quite smooth, except just at the base in front. They are a dull nankeen colour with broken brown bars. The lip is much paler, with broken rose-coloured streaks, rounded at the point and turned inwards at the base, so as to resemble a slipper. When flattened out it has a slightly repand rhomboidal outline ; on its upper side it is hairy, and bears in the centre a conspicuous 3-toothed fleshy appendage. The pollen masses are in two pairs, each placed upon a short strap, which connects it with a very narrow crescent-shaped gland. It is no doubt the form of this gland which has led to the opinion, that the plant is a Scuticaria, the two straps, which shrink up after a few days, having been overlooked.

333. BERBERIS TRIFOLIATA. *Hartweg.* A beautiful half-hardy evergreen shrub, with variegated glaucous spiny leaves. Flowers small, pure yellow. Native of Mexico. (Fig. 168.)

The charming foliage of this shrub renders it one of the most valuable of the species of this interesting genus, even although it requires some shelter in our severer winters. To the following account, in the *Botanical Register*, we find nothing to add :—

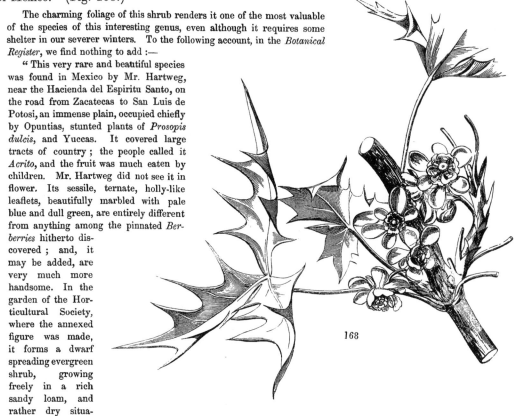

168

" This very rare and beautiful species was found in Mexico by Mr. Hartweg, near the Hacienda del Espiritu Santo, on the road from Zacatecas to San Luis de Potosi, an immense plain, occupied chiefly by Opuntias, stunted plants of *Prosopis dulcis*, and Yuccas. It covered large tracts of country ; the people called it *Acrito*, and the fruit was much eaten by children. Mr. Hartweg did not see it in flower. Its sessile, ternate, holly-like leaflets, beautifully marbled with pale blue and dull green, are entirely different from anything among the pinnated *Berberries* hitherto discovered ; and, it may be added, are very much more handsome. In the garden of the Horticultural Society, where the annexed figure was made, it forms a dwarf spreading evergreen shrub, growing freely in a rich sandy loam, and rather dry situation. It has stood two winters, planted against a south wall, and seems to be about as hardy as *Berberis fascicularis*. It may be increased in various ways ; by layering, by cuttings, or by seeds, but when the kind is rare, like the present, grafting is the most certain and safest way ; the grafting may be performed in the ordinary way, in March or April, and the best stock for working it upon is *Berberis aquifolium*. When grafted it should be placed in a cold pit or frame, kept close and rather damp. It flowers freely in April and May."

334. DOMBEYA MOLLIS. *Hooker.* (*aliàs* Astrapæa mollis *Hortulan.*) A coarse hot-house tree, with fragrant pink flowers. Belongs to Byttneriads. Native of the Isle of France. Blossomed at Kew.

" The largest of our Dombeyas, attaining, in our Palm-stove, a height of thirty feet, with a large spreading head of branches. It is an undescribed species (though nearest, perhaps, to *D. triumfettæfolia*, Bojerin *Ann. Sc. Nat.* vol. xviii. p. 191) and was received at Kew many years ago from France, under the name of *Astrapæa mollis*. The species is remarkable for its large, soft, and compactly tomentose leaves, and the dense capitate umbels of small rose-coloured flowers with narrow petals. It flowers in March, and the scent resembles that of Hawthorn. A tree, much branched at the top, spreading; young petioles, leaves, peduncles, and calyces everywhere clothed with dense stellated down, quite soft to the touch. Leaves on terete petioles, often a foot long, themselves nearly as long, cordate with a deep sinus at the base, three-lobed, the lobes very much acuminated and straight (not diverging), everywhere sharply serrated, five- to seven-nerved. Stipules moderately large, ovato-acuminate. Peduncles six to eight inches long, rather stout, erect, two or three times forked at the apex; each branch bearing a capitate umbel of pale rose-coloured flowers. Calyx of five, oblong, much acuminated sepals. Petals five, lanceolate, acuminate, falcate, but somewhat uncinate at the apex. Filaments of the stamens united into an urceolate tube. Anthers fifteen, oblong. Sterile filaments linear, subpetaloid, thrice as long as the fertile ones. Ovary globose, stellato-hirsute. Style with five linear stigmas."—*Bot. Mag.*, t. 4578.

335. BLETIA PATULA. *Hooker.* A handsome terrestrial tuberous Orchid, with rich purple flowers. Native of Cuba and Hayti. Blossoms in May. (Fig. 169, a reduced sketch; *a*, a flower in front of the natural size; *b*, the lip spread open and a little magnified.)

For specimens of this we are indebted to the Earl of Derby, with whom it blossomed at Knowsley in May last. It was imported from Hayti, whence we also received it from Mr. Charles Mackenzie; it already produces a flower-stem three feet high, with a promise of greater vigour. Mr. Linden also found what seems to be the same species on the sandy hills of Yatera, in Cuba, in May 1844, with large bright purple flowers, lanceolate leaves, a foot and a half long; very strong, roundish oval pseudo-bulbs, and a stem a foot and a half high. In general habit this is not unlike the common *Bl. verecunda*; it is still more like *B. Shepherdii*: its lip is however in no degree three-lobed; although, from the manner in which it is folded on each side of the end, it looks as if it were so. This peculiarity is well represented in the *Botanical Magazine*, t. 3518. The true form of the lip is an exact oblong, as in our cut, with a very short stalk at one end, and a deep notch at the other. It has a thin texture, is much plaited, and has along the middle from five to seven white parallel crests, which are interrupted here and there, and end abruptly below the end of the lip.

336. ROGIERA VERSICOLOR. (*aliàs* Rondeletia versicolor *Hooker.*) A stove plant with yellow and pink thyrses of flowers. Belongs to Cinchonads. Native of Central America.

"Sent by Mr. Seemann from Boqueta, Veraguas, Central America, to the Royal Gardens of Kew in 1838.

A handsome stove shrub, especially when its copious cymes of dense flowers are in perfection (March and April), and which are remarkable for the play of colours: the tube is yellow; the limb in bud deep rose-colour, changing when they expand to pale rose and then to white, with a yellow disc, and having a two-lobed green spot in the centre from the colour of the stigmas, which protrude a little beyond the mouth. It does not correspond with any of the many species now described of this genus; its nearest affinity is, perhaps, with *R. cordata* Benth. (*Rogeria* Planch. and Henfrey) from Guatemala, but that is nearly glabrous, and has sessile leaves, broad and cordate at the base. A moderate-sized shrub, with "very bitter bark." Branches obscurely four-sided, but compressed, younger ones and young leaves quite silky and shining. Leaves large, deep green, soft and submembranaceous when fresh, more hard and almost coriaceous when dry, ovate, acuminate, very obtuse or subcordate at the base, above in the adult foliage glabrous or nearly so, beneath and on the petioles (half an inch long) pubescenti-tomentose, paler in colour, veins pinnated, prominent, beneath, a good deal reticulated, the reticulation most distinct in the dry state. Stipules deciduous from the older leaves, broad ovate, spreading, membranaceo-herbaceous, downy. Panicle downy, trichotomously divided and bearing numerous flowers, so as to form a more or less dense cyme, everywhere very downy, even the outside of the corollas. Calyx-tube small, globose: teeth five, small. Corolla hypocrateriform; the limb of five, spreading, rather wavy lobes, silky in the disc. Stamens quite included. Style a little exserted. Stigma two-lobed."—*Bot. Mag.*, t. 4579.

This is very near *R. Menechma* figured above at p. 41., no. 300, but its flowers are much larger with blunter lobes. Perhaps it is the same plant, better grown.

170

b

337. CALCEOLARIA TETRAGONA. *Bentham.* A broad-leaved greenhouse shrub, with loose corymbs of large pale yellow flowers. Native of Peru. Belongs to Linariads. (Fig. 170, *a*, natural size of flowers; *b*, a diminished figure of a branch.)

This was exhibited by Messrs. Veitch, at the last great exhibition of the Horticultural Society. It forms a compact evergreen bush, with pale green, broad, oblong, blunt, entire leaves, from three to four inches long; which, in a wild state, are frequently (always?) covered with a glutinous exudation. The flowers are among the largest in the genus, pale yellow, with a very large yellowish-green calyx, consisting of blunt, spreading, oblong sepals. In habit it is wholly distinct from all those previously in cultivation, and may probably become, in the hands of skilful hybridizers, the parent of an entirely new race of Calceolarias. It seems to be a true shrub; the foliage is much better than that of other garden species, and the large flowers only want brilliancy and gay marking to be very beautiful objects.

338. SEDUM KAMTCHATICUM. *Fischer and Meyer, Ind. Seminum in Horto Petropolitano; Walpers' Repertorium,* ii. 262. Received from Dr. Fischer, in June, 1844, and said to have been collected by Dr. Schrenk on the Chinese limits of the South of Soongaria.

a

This is a handsome herbaceous plant, with bright yellow flowers like those of *Sedum Aizoon,* which it much resembles in habit. The leaves are obovate and toothed at the upper half only, but they narrow in a wedge-shaped manner to the base. They are red edged, and the stem has also a strong stain of that colour; most of them are alternate, a very few only near the summit being opposite to each other. It is a hardy perennial, requiring a light soil and dry situation. It is easily increased by cuttings any time during the summer or autumn, and flowers from June to August. It proves to be a fine showy plant for Rockwork, where it blooms freely and remains long in succession.—*Journal of Hort. Soc.*, vol. i.

339. ROSA FORTUNIANA. A scrambling evergreen hardy shrub, with large solitary flowers, and ternate or quinate leaves. Native of China. Introduced by the Horticultural Society. (Fig. 171.)

R. Fortuniana (Banksiana), ramis scandentibus glabris, aculeis parvis falcatis distantibus, foliolis 3-5nisve ovato-lanceolatis nitidis argutè serratis, floribus solitariis, calycis tubo hemispherico nudo sepalis ovatis indivisis.

Among the roses introduced by Mr. Fortune, for the Horticultural Society, is one which does not appear referable to any known species. It is a scrambling shrub, with slender branches, sparingly armed with small falcate prickles. The leaflets are ovate-lanceolate, finely serrate, thin, bright green, shining on both sides, and usually in threes, sometimes in fives. The stipules are small, subulate and deciduous as in the Banksian roses. The flowers grow singly on short setose peduncles; have a hemispherical naked calyx tube, and ovate undivided sepals, and are double white, with their petals loosely and irregularly arranged in a mass about three inches in diameter. That it is not a Banksian rose is proved by its solitary flowers and prickly stems; that it is no variety of *R. sinica*, is shown by its weaker habit, and the total absence of spines from its calyx-tube. Can it be a mule between the two? The plant has not much beauty, so far as the flowers are concerned, but its rapid growth, straggling habit, and evergreen leaves, render it extremely well suited for covering walls, verandahs, or rustic work in gardens.

171

340. Franciscea Calycina. *Bentham.* (*aliàs* Besleria inodora *Vellozi*; *aliàs* Franciscea confertiflora *Henfrey*.) A beautiful stove shrub, with large violet flowers. Native of Brazil. Belongs to Linariads.

"We continue the genus *Franciscea*, as sanctioned by Mr. Miers, in the fifth volume, new series of 'Annals of Natural History,' for the blue-flowered species of *Brunsfelsia*, though we fear Mr. Bentham's views of the unsoundness of the generic distinction are too true. We find the present plant figured and described by Mr. Henfrey in the 'Magazine of Botany,' under the name of *F. confertiflora*, and the only synonyme given is the *Brunsfelsia confertiflora* of Mr. Bentham, a species with which we are familiar, and of which there exists a splendid figure in Pohl's 'Plantarum Brasiliarum Icones :' but the figure and description are totally at variance with our plant. It is unquestionably the *F. (Brunsfelsia) calycina* of Bentham, figured, characteristically enough, in the 'Flora Fluminensis,' and well distinguished by the large inflated calyx and other characters. As we are indebted for our plant to Messrs. Lucombe, Pince, and Co., Exeter Nursery, who received it from Belgium, we presume that the Belgian horticulturists are answerable for anything wrong in the name, though that is not implied in the 'Magazine of Botany.' It is a most lovely species, and must soon be a great favourite with cultivators. Our garden is further indebted for a flowering plant to Messrs. Henderson, Pine Apple Place. It forms a compact bush, blossoming readily when eighteen inches high : and, like other real *Franciscas*, the flowers are at first violet-blue, then white, or nearly so. A moderate sized shrub, with terete, glabrous branches and copious evergreen foliage. Leaves alternate, on very short footstalks, nearly elliptical, entire, obtuse at the base, acute, or shortly acuminated at the point, glabrous, or with a slight degree of hairiness on the midrib beneath. Cymes few-flowered, generally terminal. Pedicels thickened, as long as the calyx. Calyx large, elongated, tubular and inflated, glabrous, five-toothed at the apex. Corolla large, rich purple, with a white ring round the mouth of the tube, soon changing to a pale purple, and then almost to white. Tube curved downwards, not much longer than the calyx : limb oblique with regard to the tube, more than two inches across, of five, broadly obovato-rotundate, horizontally spreading and waved segments. Stamens and style quite included."— *Bot. Mag.* t. 4583.

341. Vanda insignis. *Blume.* A fine stove epiphyte from Java. Flowers yellow and brown with a whitish lip streaked at the base. Introduced by John Knowles, Esq.

At last this long sought rarity has appeared. We owe to the kindness of John Knowles, Esq., of Manchester, a living specimen, which entirely confirms the accuracy of Dr. Blume's figure, and description. It is in the way of *V. tricolor* and *suavis* ; and the flowers are of the same size, with long white angular stalks.—*Linden.*

The sepals and petals are dull yellow blotched with brown. The lip is white, with a faint shade of violet in the middle, and rich deep crimson streaks at the base. The flowers have a slight and agreeable fragrance, even when cut and kept in a sitting-room.

342. Ixora javanica. *De Candolle.* (*aliàs* Pavetta Javanica *Blume*.) A beautiful orange-flowered stove plant, belonging to Cinchonads. Native of Java. Blossoms in March.

"From the collection of Messrs. Rollisson, Tooting, who imported this very charming species of *Ixora* from Java, and with whom it blossomed in March 1851. It is handsome in the rich coral colour of the branches, in the full green of its copious foliage, and in the large corymbs of orange-scarlet flowers. The *I. Javanica* of Paxton, Mag. of Bot. v. 14, p. 265, is very different from this, and not Blume's plant. A shrub, glabrous in every part, with compact branches, which are rounded, and the younger ones at least of a rich coral colour. Leaves four to five or even six inches long, between coriaceous and membranaceous, ovate-oblong, acute or acuminate, entire, penninerved, and acute or more or less attenuated at the base, where it gradually passes into a short petiole, not a quarter of an inch long. Stipules from a broad connate and therefore amplexicaul base, terminating suddenly in a long cuspidate spine-like point. Corymb terminal, large, on a long peduncle, which, as well as the trichotomous branches, are deep coral-coloured. Calyx almost turbinate, with two small bracteoles at the base : the limb of four, erect, rounded, obtuse lobes. Tube of the corolla an inch and a half long, slender filiform, red : limb an inch across, deep orange red, the lobes horizontally patent, obovato-rotundate. Anthers linear, when perfect lying at the mouth of the corolla, but very deciduous. Style as long as the tube of the corolla ; its thickened bifid stigma a little exserted."

"This, like the majority of the genus, is a showy species. Being a native of Java, it requires to be cultivated in a warm and moist stove ; and this is not only necessary in order to produce luxuriant growth, but also to prevent the plants from becoming infested with insects, to which the species of this and other allied genera are very commonly subject, and which often cannot be got rid of without making the plants look very unsightly and producing an unhealthy condition. Pits heated with fermenting stable litter or leaves, are well suited to the growth of such plants as Ixora ; the confined and moist atmosphere encourages a vigorous growth, and this, with the vapour arising from the fermenting matter, are great preventatives of the breeding of insects. The soil may consist of about one-half light loam and peat, or leaf-mould, with a small quantity of sharp sand, and care must be taken to drain it well, and, in shifting, not to over-pot it. This, like the rest of the genus, is readily increased by cuttings treated in the manner generally recommended for the propagation of hard-wooded stove plants."—*Bot. Mag.*, t. 4586.

343. DRIMIOPSIS MACULATA. A greenhouse bulbous plant, of little beauty, from the Cape of Good Hope. Flowers green and white. Belongs to Lilyworts. Introduced by the Horticultural Society. (Fig. 172.)

Among some bulbs received by the Horticultural Society from the Cape, came up the present plant, of which we can find no trace in books. From a white fleshy bulb there rise a few broad, fleshy, oblong leaves, six or eight inches long, rolled up at their base, so as to form a kind of channelled petiole ; they are abundantly clouded with dark green oblong stains upon a paler ground. The scape, which is taper, and about as long as the leaves, is terminated by a close raceme of half-closed campanulate flowers ; the lower of which are green and pendulous, the upper white and erect. Both sepals and petals are herbaceous, ovate, cucullate, concave, and united at the base ; the petals are rather shorter and broader than the sepals. The stamens are six, equal ; their filaments inserted by a broad base upon the sepals and petals ; the anthers are ovate and turned inwards. The ovary is ovate, roundish, undivided, gradually tapered into a style with a simple minute stigma ; in each of its three cells stand erect a pair of anatropal collateral ovules. It is evident from this description, that the plant cannot be a *Drimia*, to some of which, especially *Dr. lanceæfolia*, it bears a certain resemblance ; for it wants the petaloid spreading petals of that genus. Neither can it be a *Lachenalia*, because of its twin collateral ovules, and herbaceous perianth. We therefore propose it as a new genus, with the following character :—

DRIMIOPSIS. *Perianthium* herbaceum campanulatum subæquale. *Stamina* æqualia, epipetala. *Ovarium* in stylum attenuatum ; *ovula* gemina, collateralia—Herbæ *bulbosæ*, foliis *succulentis*, scapo racemoso, comá destituto.

344. WALLICHIA DENSIFLORA. *Martius.* (*aliàs* Wallichia oblongifolia *Griffith.*) A dwarf stemless tropical Palm, requiring a stove. Native of Assam. Flowered at Kew.

" Dr. Hooker remarks, it is common in damp forests at the foot of the Eastern Himalaya, extending at least as far west as Kamaon, where Dr. Thomson found it at an elevation of about 2000 feet above the level of the sea. It is a very elegant Palm, and very beautiful when in fructification. The male and female spadices appear on the same plant, and arise from among a tuft of strong coarse fibres : the former enveloped in large imbricated spathas of a dark purple, streaked with yellow : these separate, and then a dense cluster of male spadices appear, of a nearly white colour. The male spadix is a compound spike, with violet-coloured ovaries. Such a plant is well suited to commemorate Dr. Wallich's labours in the field of science. His extended knowledge and his splendid works on Indian Botany, his liberal contributions to Kew and to every celebrated garden in Europe and the Colonies, and his generous and encouraging bearing to every student of plants, justly entitle him to a name among the ' Princes of the Vegetable Kingdom :' a name, too, given by his predecessor in the Directorship of the Calcutta Garden, Dr. Roxburgh."

" It is seldom that we have an opportunity of offering remarks on the cultivation of Palms : this may in part be attributed to the want of show in their flowers, and the general loftiness of growth of the majority of the family. But the species figured may be viewed as an exception, for it is not only a dwarf or stemless Palm, but its large bunch of male flowers is conspicuous on account of its singular-coloured spatha. Being a native of India, it requires the heat of a tropical stove, and grows freely in any kind of light garden-soil not retentive of water. The plant from which the drawing was made was introduced into the

Royal Gardens some years ago, being then a small plant. As it increased in size and filled the pot with roots, it was duly shifted into larger pots, and ultimately into a plant-box two feet square, where it flowered, in the Palm-house. Although it produced both sexes of flowers, it did not, however, perfect its seeds. It may be increased by separating the suckers, but this must be done gradually, so as to allow the suckers time to have sufficient roots before they are quite separated from the plant."—*Bot. Mag.* t. 4584.

345. ACACIA VISCIDULA. *Bentham.* A handsome erect greenhouse shrub, with balls of deep yellow flowers in March and April. Native of New South Wales. (Fig. 173.)

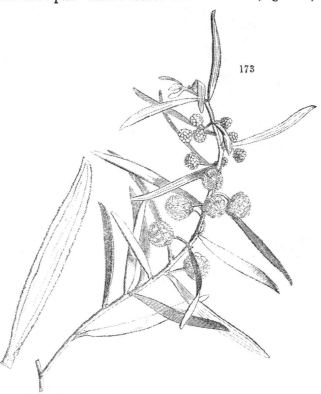

173

This plant is one of the most useful of the New Holland Acacias, not growing to a large size, and flowering profusely during all the spring. Frazer found it on the banks of the Lachlan ; and Sir Thomas Mitchell, in September, scarcely in flower, at the base of sandstone mountains, in the subtropical parts of New Holland, where it formed a tree 12 feet high. Its leaves and branches are covered with a glutinous substance, which, when dry, cracks and gives the edges of the leaves and the angles of the branches a broken appearance. In our gardens this passes under the name of *A. ixiophylla*, a very closely allied species, with short spikes of, not solitary, flower-heads, and leaves three or four times as broad. In this the flower-heads often grow in pairs, but they are not united by any common peduncle.

PLATE 52.

L. Constans del. & Zinc.

Printed by C.F.Cheffins. London.

[Plate 52.]

THE VARIOUS-LEAVED LABICHEA.

(LABICHEA DIVERSIFOLIA.)

———◆———

A Greenhouse Shrub, from SWAN RIVER, *belonging to the Order of* LEGUMINOUS PLANTS.

Specific Character.

THE VARIOUS-LEAVED LABICHEA. Leaves un-equally digitate, sessile ; the leaflets linear-lanceolate, spiny-pointed, thick-edged, smooth, those at the sides many times smaller than the middle one. Racemes few-flowered, much shorter than the leaves. Calyx and corolla each in four parts. One anther much longer than the other, with a single pore, the other with two pores.

LABICHEA *DIVERSIFOLIA ;* foliis impari-digitatis sessilibus foliolis lineari-lanceolatis spinoso-mucronatis marginatis sessilibus glabris lateralibus pluriès minori-bus, racemis paucifloris foliis multò brevioribus, calyce corollâque tetrameris, antherâ alterâ duplò longiore uniporosâ alterâ biporosâ.

Labichea diversifolia ; *Meisner in Plant. Preiss.*, i. 23.

THIS curious Leguminous genus consists of shrubs with spiny digitate leaves, of which the lateral leaflets are often very much smaller than the central one. They have short clusters of axillary yellow flowers, not unlike those of Cassia, though materially different in structure. In the original species the usual number (5) that occurs in the flowers of Leguminous plants is preserved in the calyx and corolla ; while at the same time the stamens are reduced to 2, one of whose anthers opens by 2 pores, and the other, which is much longer, opens by one. In this, on the contrary, we have only 4 sepals, and 4 petals, while the stamens remain as in the original species.

The theoretical structure of the flower in this case appears to be this : the two dorsal sepals unite into one, as is indicated by a middle line, which passes through it from the base to the apex ; this brings one of the petals, in appearance, opposite to the dorsal sepal, although really alternate with the two sepals which form it, and at the same time throws all the other petals out of their places. The number 4 in the petals is undoubtedly owing to abortion, for it is not unusual to find a fifth petal, in the form of a subulate process at the place where a star is shown in the accompanying diagram, and where one also would be if the corolla were papilionaceous. The two stamens, unequal as they are, appear to belong to the fifth or dorsal petal.

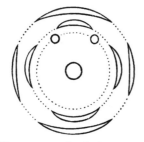
Diagram representing the flower as it is.

Diagram representing the theoretical structure of the flower.

As the memoranda which we made when examining the flowers of this plant do not coincide with those of others, we now produce them literally. " Sepals 4, of which 2 are anterior and posterior, leafy, concave, acuminate, involute, the other two membranous. Stamens 2, in front of the upper petal, unequal; one horn-shaped, opening by a simple pore; the other oblong, shorter, opening by a double pore, the one as exactly crossing the other as a soldier's crossbelts. The dorsal sepal has a fine line passing through its middle from top to bottom."

These plants are all Australian, chiefly inhabiting the western coast, but also met with in the region north of Sidney, where Sir Thomas Mitchell found two species. That now figured was seen by Preiss among fragments of quartz rocks on the west side of Mount Bakewell, and in rocky places near the river Canning in the Darling range, on the west side of New Holland.

The accompanying figure was made in the Nursery of Mr. Glendinning, of Turnham Green, in April last.

PLATE 53.

L.Constans del.& Zinc.

Printed by C.F.Cheffins,London.

[PLATE 53.]

THE LILAC THYRSACANTH.

(THYRSACANTHUS LILACINUS.)

———◦———

A Hothouse Shrub, from TROPICAL AMERICA, *belonging to the Natural Order of* ACANTHADS.

═══════════════════

Specific Character.

THE LILAC THYRSACANTH. Leaves oblong, stalked, wrinkled, acute, downy on the under side, as are the branches. Inflorescence very erect, close, compact, naked, with a downy axis. Sepals pubescent, with very fine long points. Corolla smooth ; its lobes glandular inside ; the upper lip 2-lobed, the lower closely reflexed. Sterile stamens awl-shaped at the point.	THYRSACANTHUS *LILACINUS;* foliis oblongis petiolatis rugosis acutis subtus ramisque tomentosis, inflorescentiâ strictâ nudâ, axi tomentosâ, sepalis pubescentibus setaceo-acuminatis, corollæ glabræ laciniis intus glandulosis, labio supremo bilobo inferiore arctè deflexo, staminibus sterilibus apice subulatis.

———

Justicia lilacina of the Gardens.

═══════════════════

" SEPALS 5, subulate, equal, pilose. Corolla funnel-shaped, 4-lobed, inflated at the base, then contracted into a slender tube ; its segments glandular inside ; three oval, nearly equal, reflexed, the fourth erect, oblong, 2-lobed, ciliated. Stamens 2 fertile, lying beneath the fourth petal ; 2 others subulate, sterile, enclosed in the tube. Anthers fleshy, linear, 2-celled, with parallel nearly equal lobes. Ovary linear-oblong, seated on a roundish torus, 2-celled, many-seeded. Stigma minute, 2-toothed."

Such were the memoranda made by us when this plant was under examination. They rendered it evident that it was a species of *Thyrsacanthus,* a genus carved by Nees von Esenbeck out of the farrago called Justicia by previous writers, of which the common white-flowered *Justicia nitida* is the most familiar example. Among the species enumerated by this writer, one more especially, his *T. bracteolatus,* seemed to agree in character with the plant before us. That species had been previously called *Justicia bracteolata* by Jacquin, who figured it under the name in his Icones, vol. ii., t. 205. But *J. bracteolata* is evidently identical with Nees von Esenbeck's *Th. Lemaireanus,* of which a good figure is to be found in the Flore des Serres, t. 240, under the name of *Eranthemum coccineum ;* * and in this plant the flowers are red, the lower lip spreading not reflexed, and the sterile stamens spathulate not subulate. With *Thyrsacanthus bracteolatus,* aliàs *Lemaireanus,* aliàs *Justicia bracteolata,* aliàs *Eranthemum coccineum,* it is clear that our plant is not to be confounded.

Nor can we find any trace of the species elsewhere. Its garden name of *Justicia lilacina* affords no clue; and we, therefore, conclude it to be one of the many Acanthads which have not hitherto found a niche in the halls of Systematic Botany.

Our drawing of it was made in March last, in the Garden of the Horticultural Society, where it had been flowering in a stove all through the winter, and where its pretty lilac flowers, so rare a winter colour, had proved a great decoration. Its native country is unknown; but we presume it must belong to some region of tropical America.

* Other *aliàses* of this plant are *Aphelandra longiracemosa, Aphelandra longiscapa, Salpingantha coccinea,* and *Justicia longiracemosa*—all Garden names.

PLATE. 54.

L. Constans. del & Zinc.

Printed by C.F.Cheffins, London.

[PLATE 54.]

THE CARMINE TRICHOPIL.

(TRICHOPILIA COCCINEA.)

———◆———

A beautiful Epiphyte, from CENTRAL AMERICA, *belonging to the Natural Order of* ORCHIDS.

Specific Character.

THE CARMINE TRICHOPIL. Pseudobulbs oblong, narrow, compressed, furrowed, one-leaved. Leaves lanceolate, flat, somewhat cordate at the base, acuminate and recurved at the point. Peduncles one-flowered. Petals linear-lanceolate, acuminate, twisted once. Lip 4-lobed, closely rolled up at the base; its divisions rounded, convex, plane. Hood 3-lobed with fringed nearly equal divisions.

TRICHOPILIA *COCCINEA ;* pseudobulbis oblongis angustis compressis sulcatis monophyllis, foliis lanceolatis planis basi subcordatis apice acuminatis recurvis, pedunculis unifloris, petalis lineari-lanceolatis acuminatis semel tortis, labello quadrilobo lobis rotundatis convexis planís basi arctè convoluto, cuculli trilobi laciniis fimbriatis subæqualibus.

Trichopilia coccinea : *Warczewicz in his correspondence and throughout the English auctions and gardens in* 1849 *and* 1850 : *aliàs* Trichopilia marginata : *Gard. Mag. of Bot., July* 1851, *with a figure.*

THIS beautiful species of Trichopil was found in Central America by Mr. Warczewicz (not by Linden as has been stated), by whom drawings and living plants were sent to England in 1849, under the name of *T. coccinea,* by which it was publicly sold, and has since been universally known. We cannot therefore subscribe to Mr. Henfrey's alteration of the name to *T. marginata,* either in justice to the zealous and ill-requited traveller who found it, or in the interest of science, which suffers seriously in public estimation by the needless changes in names made by writers on Natural History subjects.

The usual colour of the flower is a deep rich carmine, with a narrow edge of white; but it appears from the figure above referred to, that the rich colour is sometimes confined to the lining of the tube, the whole of the expanded limb being white. The drawing from which our plate was prepared was made in the Garden of the Horticultural Society last May; the colour is less intense than in Mr. Warczewicz's unpublished drawings, but we suspect will continue to improve as the plants become more and more healthy. It already answers very nearly to the term " coccineus " or pure carmine colour, applied to it by its discoverer.

It is not to the Sweet Trichopil (our plate 11) that we must look for resemblance to this species, for the whole form, texture, and aspect of that plant are different. It is to the original Corkscrew Trichopil (*T. tortilis*) that it approaches nearly, differing principally in its larger and rich carmine flowers, slightly twisted sepals and petals, and the equal size of the fringed lobes of the anther-hood. In foliage and pseudo-bulbs the two are so much alike, that one might be taken for a more vigorous specimen of the other.

Now that we have three well-ascertained species before us, it may be as well to point out the differential characters of what are known, thus :—

T. tortilis Lindley. Pseudobulbs narrow, compressed, furrowed. Leaves lanceolate, plane, slightly cordate. Petals twice twisted, brown and yellow. Lip even, flat, white with crimson spots.

T. coccinea Warczewicz. Pseudobulbs narrow, compressed, furrowed. Leaves lanceolate, plane, slightly cordate. Petals once twisted, brownish and yellow. Lip even, flat, carmine with a white border.

T. suavis Lindley. Pseudobulbs thin, orbicular. Leaves broad, oblong, undulated. Petals not twisted, white dashed with pink. Lip very thin and wavy, crisp, white with rose-coloured blotches.

T. Galeottiana Richard and Galeotti, Orch. Mex. t. 31 ined. Pseudobulbs terete, stem-like. Sepals and petals not twisted. Flowers very large, yellow.

This last species is at present not further described.

346. Cerasus nepalensis. *Seringe.* A hardy deciduous tree, with white flowers, from Nepal. Introduced by the Horticultural Society. Blossoms in June. (Fig. 174.)

This is very like our common Bird-cherry, and must be regarded as its Indian representative. The leaves are

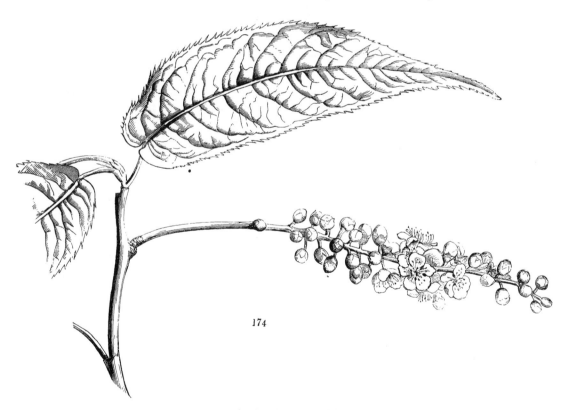

174

cordate at the base, very glaucous underneath, where also the veins are remarkably shaggy. The peduncles and pedicels are alike downy. The flowers are smaller than in the European species. We suppose there can be no doubt about this being the *Cerasus nepalensis* of Seringe, notwithstanding the apparent errors and material discrepancies in his specific character, for there are but two Bird-cherries in the North of India, namely this and *C. undulata,* much better called *capricida* by Dr. Wallich, and the latter is a very different plant. It is much to be regretted that Mons. Seringe should not have been aware of Dr. Wallich's catalogue names, when he published the Rosaceæ in De Candolle's Prodromus, in which case the present plant would have borne the name of *C. glaucifolia*, so much more appropriate than *C. nepalensis*.

Dr. Wallich states that the present plant is found in both Nepal and Kamaon.

347. Pyxidanthera barbulata. *Michaux.* (*aliàs* Diapensia barbulata *Elliott; aliàs* Diapensia cuneifolia *Salisbury.*) A charming prostrate shrub, with small pink flowers. Native of the United States. Belongs to the Order of Diapensiads.

Early in the month of May I was gratified on the arrival of the Royal Mail Steamer from New York, with tufts of this charming little plant sent me by Mr. Evans of Radnor, Delaware, gathered in the pine-barrens of New Jersey, as fresh and as full of perfect flowers as if that day removed from the native soil. These have given me the means of publishing the accompanying figure, of which, as far as we know, no other representation has been given than the very indifferent one of Michaux. The genus we think correctly distinguished from *Diapensia* by the aristate anthers and few-seeded capsules and habit. It is more difficult to determine the place of this little family. It clearly belongs to the "*Corollifloræ*," yet De Candolle has hitherto passed it by. Brown removes it from *Convolvulaceæ*, where Jussieu was inclined to place it. Salisbury referred it to *Ericaceæ*, but apparently with little reason; and Endlicher says of it, "*Ericaceis* affinis." Dr. Lindley places it between *Loganiaceæ* and *Stilbaceæ.*—If it should prove easy of cultivation *Pyxidanthera* would make a charming rock-plant : the rose-coloured buds are as pretty, nestling among the copious foliage, as the fully expanded white flowers. A small, tufted, procumbent, creeping, and wide-spreading shrub, having a long tap-root in the centre of the tuft : branches terete, slender, younger ones woolly. Leaves alternate, cuneato-oblong, very acute, almost aristate, the young ones woolly at their base within, and hence the specific name of "*barbulata.*" That character disappears in the older portions of the plant. Flowers solitary sessile, from little branches with rosulate leaves. Calyx of five, concave, reddish sepals, as long as the tube of the corolla. Corolla monopetalous, white : tube short : limb of five, rounded-cuneate, spreading, slightly crenated lobes. Stamens in the sinuses of the corolla. Filaments broad, white, almost petalloid, bearing a drooping yellow anther of two almost globose lobes, opening transversely, and bearing an awn on the lower valve. Ovary ovate, with a thickened ring at the base, three-celled, few-seeded (four or five in each cell) attached to a central placenta. Style as long as the tube of the corolla. Stigma of three small spreading rays. We have several times received from the United States flowering tufts of this very small shrub; but although they have been placed under different kinds of treatment, both in the open air and under protection, we have not yet succeeded in keeping them long alive. Dr. Asa Gray informs us that the shrub grows in the warm "pine-barrens" of New Jersey, in low but not wet places, generally on little knolls, fully exposed to the sun, in a soil of pure sand mixed with vegetable mould. We have examined the soil in which it grows, which we find no difficulty in imitating, and by attention the proper degree of moisture and temperature can be maintained; but as it has not thriven under our care, we infer that the want of success is owing to some peculiarity in its nature, together with the difference between the climate of this country and that of its native locality. One thing to be noticed is that our imported plants have certainly been very old, having (comparatively) long wiry roots like the old roots of a heath. It is probable that our cultivation might meet with better success if young plants could be procured, either from cuttings or from seeds.—*Bot. Mag.*, t. 4592.

348. Dendrobium villosulum. *Wallich.* A handsome Indian epiphyte, with rich orange-coloured flowers and rough stems. Flowers in June. Introduced by the Honourable the East India Company. (Fig. 175.)

When Dr. Wallich's vast accumulation of botanical records was, by the great and wise liberality of the East India Company, dispersed through all civilised lands, this plant was a mere fragment, without any one thing to show that it was a Dendrobe at all, beyond its peculiar habit. Recently (May, 1851) Mr. Loddiges has produced it in flower; and we are

now enabled to show that it is not only really a Dendrobium, but one of a most remarkable and brilliant nature. It was imported from Tillicherry.

At the end of long slender stems, clothed with short black hairs, appear rich orange-coloured flowers in pairs. Their sepals and petals are linear, concave, obtuse, curved like so many horns, the petals being broader at the base than the sepals, and the lateral sepals forming a very short obtuse chin. The lip is linear-lanceolate, 3-lobed, the lateral lobes being extremely short, with three wavy elevated lines running through the middle lobe from end to end. The plant is near Wallich's *Dendrobium angulatum*, with which it may be contrasted by the following character :—

D. villosulum (Endendrobium) caule erecto nigro-villoso, foliis linearibus acutè et obliquè bilobis, pedunculis bifloris, sepalis petalisque acuminatis recurvis obtusis lateralibus in mentum breve cornutum connatis, labello lineari-lanceolato trilobo 3-lamellato lobis lateralibus nanis.

349. Eremostachys laciniata. *Bunge.*
A fine showy hardy perennial from the Caucasus, with large yellow flowers. Belongs to Labiates. (Fig. 176.)

Radical leaves deeply pinnatifid with oblong-lanceolate or linear lacerated segments. Flowering stem 4-6 feet high, bearing whorls of large yellow flowers, seated in shaggy white calyxes, and supported by sessile blunt broad many-lobed green bracts. It is a common inhabitant of the eastern side of Caucasus, and of the adjoining countries, where it is found on dry hills. Its great fleshy roots are evidently adapted to such situations only. In a wild state it is not half the size of the cultivated plant, nor are its leaves half the breadth : but at the same time the flowers seem to be larger and more conspicuous. The plant appears intended by nature to resist even a Persian summer. The accompanying figure was made in April last in the Garden of the Horticultural Society, where it had been raised from seeds received from the Imperial Botanic Garden at Petersburgh. It proves to be a hardy perennial, with large spindle-shaped roots, and a stem from four to six feet in height. It is rather difficult to cultivate in the open border on account of the large fleshy roots suffering in winter from excess of moisture, but it succeeds tolerably well if grown in pots during the winter, and kept nearly dry in a cold pit or frame. It thrives in a light rich sandy loam, and flowers in May or June. It is only to be increased by seeds, and the plants are two or three years before they bloom. Care must be taken that, in potting or planting, one-third of the fleshy roots are left above ground, otherwise they soon perish.

350. Pitcairnia Montalbensis. *Linden.*
A handsome scarlet-flowered hot-house perennial, belonging to the Natural Order of Bromeliads. Native of New Grenada. Introduced by Mr. Linden.

In the *Allgemeine Gartenzeitung*, May 3, 1851, this fine plant is said to be of Mexican origin, having been discovered by Mr. Linden's collectors Funk and

176

Schlim. But as those travellers were employed in N. Grenada, the statement seems to be a mistake. It has long linear-lanceolate leaves, which are smooth on both sides and shining, and spiny-toothed at the base ; the scape is as long as the leaves, covered with a fine wool as well as the slender bracts. The spike is about 3 inches long, the corolla 1½ to 2 inches long, and scarlet-red. It would seem to be a species of some interest to cultivators.

351. RHAMNUS CROCEUS. *Nuttall.* A half-hardy evergreen shrub, from California. Belongs to the Order of Buckthorns. Introduced by the Horticultural Society. Flowers green.

Raised from seed received from Hartweg in January, 1848, and marked "a dwarf evergreen shrub, near the sea shore, Monterey."—A small evergreen bush, first described by Mr. Nuttall, who found it in bushy hills and thickets near Monterey, and who describes it as " A much branched thorny shrub, with yellow wood; the whole plant imparted a yellow colour to water. Leaves about half an inch long, lucid, when dry of a bright yellowish-brown beneath: petioles about one line long. Fascicles 2-6 flowered: pedicels as long as the petioles. Sepals ovate, with one middle and two marginal nerves. Stamens nearly as long as the sepals. Ovary ovate. Styles often distinct below the middle. Fruit greenish or yellowish, usually (by abortion) one-seeded. Seed with a longitudinal furrow on one side."

In the garden it proves to be a neat small-leaved evergreen, which, if hardy, would be a useful shrubbery plant; but near London it is tender. It flowers in June.—*Journal of Hort. Soc.,* vol. vi.

352. EURYBIA ALPINA. A hardy evergreen shrub, from New Zealand, belonging to the Order of Composites. Flowers dirty white. Introduced by Messrs. Veitch. (Fig. 177, a diminished sketch ; 1, a cluster of flowers of the natural size.)

E. alpina (Argophyllæa) fruticosa densa, ramis angulatis subtomentosis, foliis alternis petiolatis coriaceis oblongis acutis dentatis supra glabris subtus pallidis adpressè tomentosis, capitulis densè paniculatis, involucris villosis tomentosisve.

In this instance we have a further proof of the hardiness of some of the evergreen Australian vegetation, especially in the Order of Composites. *Swammerdamia antennifera* is now becoming a common evergreen; and Messrs. Veitch produced this in full flower, or rather past flower, at the May meeting of the Horticultural Society, from the open nursery at Exeter. It forms a stout bush, with angular strong branches, and firm, leathery, evergreen leaves, from 2 to 2½ inches long, deep green on the upper side, pale and somewhat hoary beneath. They are much concealed by the large quantity of dirty white flowers, which as they go off greatly diminish the neatness of the plant, especially as the florets drop off and make way for a dirty brown pappus, which becomes very conspicuous.

We find this plant among dried specimens collected in New Zealand by Mr. Bidwill, at the elevation of 8000 feet above the sea in the northern island. He describes it as a shrub 6 feet high, and believes it to be the same as a coast plant of which he also sent home specimens. The latter has larger, thinner, longer leaves, much more tapering to the base ; but may nevertheless be only a lowland form. The species is nearly allied to *E. furfuracea,* a New Zealand species with scurfy entire leaves, and also to the New Holland *E. argophylla* or Musk Tree.

353. PITCAIRNIA EXSCAPA. *Hooker.* A handsome hot-house perennial, with crimson flowers, belonging to Bromeliads. Native of New Grenada. Introduced by Messrs. Jackson and Son.

This very curious and rather handsome *Pitcairnia* was detected, as an infant plant, among some Orchidaceæ purchased from New Grenada, by Mr. Jackson of the Kingston Nursery, Surrey. They were carefully reared, and our figure represents two of them in a flowering state. The species is remarkable for the great length of the very attenuated leaves, and no less so for the sessile and densely bracteated spike of red flowers. I can nowhere find such a species described. It belongs, as far as the structure of the flowers is concerned, to the same group as *Pitcairnia suaveolens,* Lindl., figured in Botanical Register, t. 1069, that is to say, where the petals have a certain twist, occasioning their apices to point one way, and there is, moreover, a curvature there, giving a galeated character to these petals. We possess, from New Grenada, two other stemless and scapeless (or nearly so) *Pitcairnias,* and there, too, the bracteas are mixed with black spines : but in those the spines themselves bear short spreading spines on the sides. Stemless or nearly so. A kind of pseudo-bulb is formed at the base of the plant, sheathed by the dilated, dark brown bases of the outer leaves. The leaves, therefore, may be said to spring from the root, and are, many of them, full three feet long, like those of a coarse *Carex*, linear, carinated externally and gradually attenuated into a very long narrow point, quite entire, glabrous, a part of the upper margin of the sheath being alone ciliated, rather strongly so. From the centre of these leaves appears a nearly sessile, ovate head of flowers, in part concealed by numerous bracteas, imbricating each other ; the inner ones longer, narrower, yellowish-green, glabrous, the outer brown, broader, and hairy or cobwebby : these bracteas are intermingled with a few strong, acicular, almost brown spines. Calyx quite concealed by the bracteas, yellow-green : sepals lanceolate, acuminate, hairy. Petals red, curved and galeate, bearing a notched scale at the base within. Stamens shorter than the petals. Ovary superior, trisulcate. Style elongated. Stigmas three, twisted. This plant requires a warm stove, and thrives in any kind of light open soil not retentive of moisture. Care must be taken not to water it too copiously. The old roots of this species, like those of many of its allies, after a time lose their vitality, and, by their continued increase, become a nidus of support to the succeeding young roots ; but in cultivation it is advisable occasionally to turn the plant out of the pot and divest it entirely of the old roots, at the same time cutting away the lower part of the caudex, which will also be found to be dead. The plant on being repotted will soon emit young roots, and show a more vigorous growth. It is increased by offsets, and our plant shows at this time the appearance of producing perfect seeds.—*Bot. Mag.*, t. 4591.

354. Spiræa Douglasii. *Hooker.* A very fine hardy shrub, with deep rose-coloured flowers. Native of Oregon. Belongs to Roseworts. (Fig. 178.)

This brilliant addition to our Shrubberies is one of the hardiest of the North American Flora, naturally growing as far to the Northward as the straits of St. Juan de Fuca. Douglas found it on the plains of Oregon. In general appearance it resembles the *Spiræa tomentosa* of the United States, from which it differs in the following particulars :— It grows as well, if not better, in common garden soil as in peat. It is twice as robust a plant. Its leaves are longer, narrower, serrated not crenate, and white not brown underneath. The flowers are a deeper rose colour, and therefore handsomer, and form a larger and closer panicle, which always terminates in a round extremity, and is not taper-pointed. Moreover the carpels are perfectly smooth, and not buried in long down. It is one of the best shrubs in the Garden of the Horticultural Society, where our drawing was made in July.

355. Acacia grandis. *Henfrey.* A New Holland shrub, of the Leguminous Order, from the Swan River Colony. Flowers in yellow balls, in the spring.

This seems to be in no respect different from Mr. Bentham's *A. lasiocarpa,* published years ago, as far as can be ascertained from the materials laid before the public. It may be described in popular terms as a good variety of *A. pulchella,* with larger and more copious balls of flowers.

356. Epidendrum coriifolium. *Lindley.* A greenhouse epiphyte, native of Central America. Flowers green. Introduced by the Horticultural Society.

This singular plant is, in all its parts, of a tough, thick, leathery texture, and is generally glazed, as it were, with a shining exudation. The narrow stiff leaves are blunt, about six inches long,

178

concave, wtth a sharp midrib. The spike, which is terminal, and about four inches long, consists of hard amplexicaul keeled bracts pressed close to the flowers, and forming a kind of cone before they expand. The flowers are pale green, very firm and leathery, with a broad roundish convex lip, having an elevated callosity along the middle. The lateral sepals, which are particularly thick, have a serrated keel at the back.

This is a species of no beauty, nearly related to *Ep. rigidum*, but its leaves are much longer and narrower, and the flowers three or four times as large, and extremely coriaceous. It flowers in March or April in the stove.—*Jour. of Hort. Soc.*, vol. vi.

357. ACACIA HISPIDISSIMA. *De Candolle.* (*aliàs* A. Cycnorum *Bentham.*) A handsome green-house shrub, with deep yellow clusters of flowers, and very hispid branches. A native of Swan River. Flowers at Kew in the early spring.

A Swan River plant, introduced by Mr. Drummond. There are four Acacias enumerated by Mr. Bentham as nearly allied to, and perhaps not really distinct from each other ; *A. pulchella* Brown, figured in Lodd. Bot. Cab. t. 212; *A. lasiocarpa* Benth. ; *A. hispidissima* De Cand. ; and *A. Cycnorum* Benth.,—all from the Swan River settlement. Our plant accords best with the *A. hispidissima*, except that it should have pedicellated glands on the leaves, whereas both our native and cultivated specimens are destitute of them ; in this particular agreeing with the *A. Cycnorum*, which, however, ought to have pubescent and not very hispid branches. It may thus, we think, fairly be conceded that *A. Cycnorum* and *A. hispidissima* are but varieties of each other. The present is a very handsome species, having much larger leaflets and much larger capitula of flowers than *A. pulchella*, and these flowers of a rich deep yellow colour. It is, further, much stouter and more compactly growing than that species, forming very dense masses of foliage, and equally dense globose heads of flowers.

A much-branching shrub, with angular branches, and these branches and branchlets, and peduncles too, downy and densely hispid with spreading hairs, varying much in length. Leaves copious, nearly sessile, dark green : pinnæ unijugate, bearing five to seven oblong leaflets, which are obtuse, glabrous, or ciliated. A sharp acicular reddish spine is situated at the base of the leaf, and is about half its length. From the base of the leaf also the peduncles appear generally, in pairs, shorter (usually) than the leaf, and bearing a dense golden head of numerous little flowers.

This showy *Acacia*, like most of the Australian species of that genus, requires the protection of the greenhouse. It thrives in a mixture of light loam and sandy peat-soil, and, being a free grower, is well adapted either for planting out in the conservatory border or for growing in a pot. If due attention is paid to training and stopping the leading shoots, it will soon form a neat round bushy plant, and in spring present a gay appearance when in flower. It is increased by seeds, which vegetate readily in a moderate heat.—*Bot. Mag.*, t. 4588.

358. RHAMNUS HIRSUTUS. *Wight and Arnott.* A hardy deciduous shrub from the mountains of India. Flowers green ; appearing in June. (Fig. 179.)

This shrub is thus described by Dr. Wight. " Young branches pubescent, spinescent, older ones glabrous, with a white cuticle ; leaves opposite or alternate, ovate, or oblong lanceolate, with a short sudden acumination, serrulated, membranaceous, nearly glabrous above : beneath hairy, particularly on the nerves and veins ; pedicels from the base of the young shoots, 3-6 together, pubescent, as long as the petiole : calyx 4-cleft ; petals obovate, obtuse, entire flat ; ovary 2-3 celled ; styles 2-3, connected to the middle, then diverging ; the upper part jointed with and deciduous from the persistent lower half ; fruit 2-celled ; seeds plano-convex, with a deep furrow at the base on the outer convex side. A considerable shrub, rather extensively distributed on the Neilgherry hills, but not so common on the higher ranges as lower down ; it usually presents a rather scraggy appearance. It is to be met with in flower at almost all seasons." To this we can only add that the species is extremely like *Rhamnus catharticus*, from which, however, its hairiness readily distinguishes it.

359. CORIARIA NEPALENSIS. *Wallich.* A trailing, hardy, Himalayan, deciduous bush, with clusters of brownish-red flowers. Belongs to the neighbourhood of Ochnads. (Fig. 180.)

According to Wallich this is either a shrub eight to ten feet high, or a small tree, twelve to sixteen feet high, in its native mountain valleys of Nepal and Deyra Doon. In this country it is too much injured by frost to acquire any such stature; but it is nevertheless hardy enough, sending up stout four-cornered shoots from its roots if the old stems perish. Its leaves are smooth, 3-5 nerved, oblong, acute, in opposite pairs, but placed in a distichous order. The flowers appear in May, upon leafless branches, in short imbricated drooping spikes. They consist of five ovate, imbricated, acute sepals, as many small scale-like petals, ten hypogynous stamens, and five lenticular carpels placed obliquely on a conical torus or gynobase, with five free linear spreading stigmas. According to Royle (Illustrations, p. 165) this plant has given its name *Mussooree* to the Nepal province now so called, where it is most abundant at an elevation of from 5000 to 7000 feet.

Its succulent fruits are, he says, frequently eaten in the hills, though those of the common Spanish species (*C. myrti-folia*) are considered poisonous, when taken in any quantity. Griffith, who found it on the Bootan Mountains, merely says that it is a small bush (*fruticulus*) with long weak branches, crimson anthers, and stigmas looking something like a *Xanthoxylum* which he calls " *Geeree nuddee.*" It occurred at the height of from 3400 to 6000 feet. His remark about its resemblance to *Xanthoxylum* is curious, and assists in establishing the claim of *Coriaria* to a place in the *Rutal* alliance, where we have formerly stationed it.

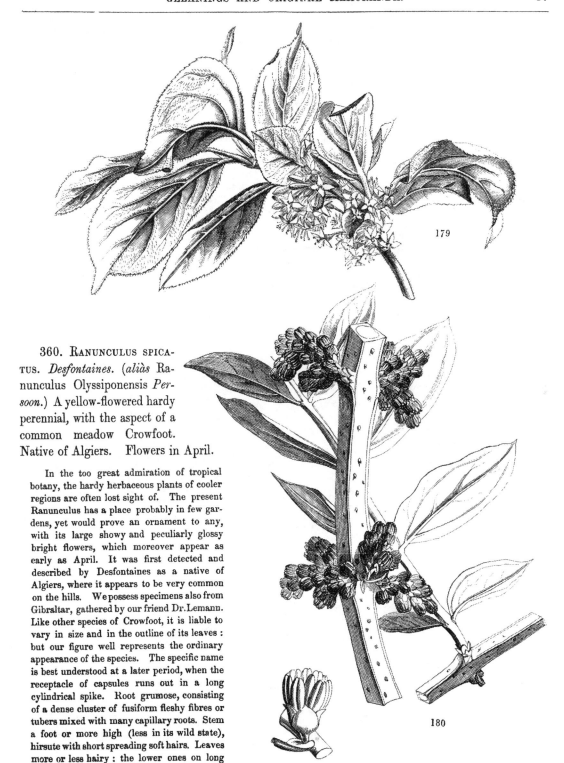

360. Ranunculus spica-tus. *Desfontaines.* (*aliàs* Ranunculus Olyssiponensis *Persoon.*) A yellow-flowered hardy perennial, with the aspect of a common meadow Crowfoot. Native of Algiers. Flowers in April.

In the too great admiration of tropical botany, the hardy herbaceous plants of cooler regions are often lost sight of. The present Ranunculus has a place probably in few gardens, yet would prove an ornament to any, with its large showy and peculiarly glossy bright flowers, which moreover appear as early as April. It was first detected and described by Desfontaines as a native of Algiers, where it appears to be very common on the hills. We possess specimens also from Gibraltar, gathered by our friend Dr. Lemann. Like other species of Crowfoot, it is liable to vary in size and in the outline of its leaves : but our figure well represents the ordinary appearance of the species. The specific name is best understood at a later period, when the receptacle of capsules runs out in a long cylindrical spike. Root grumose, consisting of a dense cluster of fusiform fleshy fibres or tubers mixed with many capillary roots. Stem a foot or more high (less in its wild state), hirsute with short spreading soft hairs. Leaves more or less hairy : the lower ones on long petioles, reniformi-orbicular, three, the lowermost five-lobed ; lobes cuneate, generally again three-lobed and incised or toothed ; upper ones nearly sessile, wedge-shaped, deeply three-lobed, and incised, the lobes linear-cuneate. Flowers one to six upon a stem, on hairy, terete peduncles. Calyx of five ovate-oblong spreading hairy herbaceous sepals. Corolla two inches broad, in cultivation, of five large, oblong, very glossy yellow spreading petals, with flabelliform, orange-coloured spots at the base. Stamens numerous, surrounding an oblong head of young carpels, which eventually lengthens into a narrow cylindrical spike.—*Bot. Mag.* t. 4585.

361. BERBERIS UMBELLATA. *Wallich.* (*aliàs* B. angulosa *and* gracilis *of Gardens.*) A handsome hardy evergreen bush, with pale yellow flowers, appearing in May. Native of the Himalayan mountains. (Fig. 181.)

Dr. Wallich's collectors appear to have first discovered this plant in Kamaon and Gossain Than. For its introduction to our gardens we are indebted to the East India Company. It is a hardy bush, about 4 feet high, with a spreading manner of growth, pale brown, angular branches, slender 3-parted spines, and very narrow, bluish-green leaves, strikingly glaucous beneath ; on an average they are 1¾ inch long by ⅜ wide ; sometimes they are perfectly entire, in which state they are represented in the " Botanical Register ;" but they are more commonly furnished with a strong, marginal, spiny tooth or two, and sometimes with many. (Can this state be the *B. ceratophylla* of G. Don ?) The flowers are pale yellow, in drooping, narrow racemes, and are succeeded by an abundance of oblong, purplish fruits. The species is very pretty, in consequence of its graceful manner of growth. It is best suited for growing among rough places, such as heaps of rockwork, where its spreading way of branching can best be seen. It is not, however, a good evergreen, the leaves being too thin and pallid.—*Journ. of Hort. Soc.*, vol. v.

181

L.Constans del. & Zinc.

Printed by C.F.Cheffins London.

[PLATE 55.]

THE MANY-COLOURED COLLINSIA.

(COLLINSIA MULTICOLOR.)

———◆———

A beautiful hardy Annual, from California, belonging to the Natural Order of LINARIADS.

═══════════════

Specific Character.

THE MANY-COLOURED COLLINSIA. Stem tall, downy. Floral leaves smooth underneath, the lowest cordate, stem-clasping, ovate-lanceolate, bluntly serrated, the middle ones linear, blunt and entire, longer than the flowers, the uppermost abortive. Pedicels with scarcely any glands, as long as the calyx which is still more glandless, with linear-lanceolate three-ribbed lobes much shorter than the corolla.

COLLINSIA *MULTICOLOR;* caule elato pubescente, foliis floralibus subtùs glaberrimis inferioribus cordatis amplexi-caulibus ovato-lanceolatis obtusè serratis intermediis linearibus obtusis integerrimis floribus longioribus supremis abortientibus, pedicellis parcè glandulosis calyci æqua-libus, calycis laciniis vix glandulosis lineari-lanceolatis 3-nerviis corollâ multò brevioribus.

═══════════════

SEEDS of this beautiful annual were received from their collector in California by Messrs. Veitch of Exeter, from whom we obtained specimens in May last.

Like *Collinsia bicolor* it grows from 1 to 1½ feet high, and is loaded with a profusion of gay flowers; but it is very much handsomer on account of the rich purple tint of its long floral leaves, and the gay markings of its party-coloured flowers. The middle boat-shaped lobe of the lower lip of the corolla is crimson, the lower lip itself is lilac, and so is the upper lip except that there is a broad white spot in its middle relieved by numerous sanguine speckles.

From *C. tinctoria* this differs in its larger flowers, seated on long pedicels instead of being sessile, the result of which is a more loose and beautiful inflorescence.

In the arrangement of the flowers it agrees with *C. bicolor*, but it has scarcely any glands on its calyx, the lobes of which are linear not ovate, and its leaves are much larger, longer, and more coarsely toothed.

PLATE 56.

L.Constans del.& Zinc.

Printed by C.F.Cheffins, London.

[PLATE 56.]

THE ROSY GREVILLEA.

(GREVILLEA ROSEA.)

————◆————

A Greenhouse Shrub, from NEW HOLLAND, *belonging to the Natural Order of* PROTEADS.

═══════════════

Specific Character.

THE ROSY GREVILLEA. Leaves simple, linear-lanceolate, rolled back at the edge, pungent, rather scabrous above, covered with a silky down on the under side. Clusters terminal, few-flowered. Calyx rose-coloured, silky, twice as short as the smooth style. Stigma oblique, depressed. Ovary stalked, shaggy. Follicle oval, mucronate, tomentose.

GREVILLEA *ROSEA* (Lissostylis) ; foliis simplicibus lineari-lanceolatis margine revolutis pungentibus supra subasperis subtus pube adpressâ sericeis, fasciculis terminalibus paucifloris, calycibus roseis sericeis stylis glaberrimis duplo brevioribus, stigmate obliquo depresso, ovario stipitato villoso, folliculo ovali mucronato tomentoso.

═══════════════

THIS very pretty greenhouse shrub was sent us some months since by the Messrs. Henderson, of Pine-apple Place, with no information as to the source from which it was derived. We find, however, a specimen in our Herbarium, collected in South Australia, which we owe to the kindness of His Excellency Governor Hutt.

It forms a neat compact bush, loaded with rich rose-coloured flowers, as large as those of *Gr. punicea*, and must be regarded as one of the most useful species in cultivation. The leaves are variable in breadth; in the specimen here represented they were quite linear; in another, which forms the vignette to this article, they were three times as broad, and but little turned back at the edge; in Mr. Hutt's specimens both kinds of leaf are on the same branch. In all cases they are scabrous above and terminated by a sharp spine. The calyx is covered with close hairs on the outside, especially at the point; the ovary is stalked, shaggy but not woolly, and has a concave smooth shallow gland at the base; the style, which is twice as long as the calyx, is knee-jointed, perfectly smooth on the upper half, but downy and even hairy as the ovary is approached. The stigma is oblong, oblique, slightly concave, with four depressed radiating lines. The fruit is brittle, oval, downy, stalked, about an inch long, containing two narrow bright brown seeds, furnished with a very narrow wing which expands into a thin appendage at the point.

We are unable to trace this among any published descriptions. It evidently belongs to Brown's section Lissostylis, and might almost be taken for a pungent-leaved form of *Gr. punicea*, but the ovary of that species, and of all its allies, is perfectly smooth. There is a *Gr. aspera* from the south coast of New Holland, which we have not seen, but that plant is said to have a very short style and linear-oblong obtuse leaves. As to *Grevillea concinna*, placed by Dr. Brown next his *aspera*, it has much longer leaves, and more copious secund racemes of flowers ; its ovary is, however, shaggy, and it seems to approach this more than anything else.

The following woodcut represents the form above alluded to, in which we could discover no difference beyond the breadth and flatness of the leaves.

L. Constans del & Zinc.

Printed by C.F. Cheffins, London.

THE WHITE AND SANGUINE DENDROBE.

(DENDROBIUM ALBOSANGUINEUM.)

———◆———

A Hothouse Epiphyte, from Moulmein, belonging to the Natural Order of ORCHIDS.

═══════════════════

Specific Character.

THE WHITE AND SANGUINE DENDROBE (Euden-drobes).—Stems thick, erect. Flowers in pairs, nodding, with herbaceous scale-like bracts. Sepals linear-lanceolate, the side ones extended into a short, obtuse, rounded chin. Petals oblong, incurved, very much broader. Lip roundish-obovate, flat, retuse, apiculate, quite entire.	DENDROBIUM *ALBOSANGUINEUM* (Eudendrobium); caulibus crassis erectis, floribus geminis nutantibus bracteis herbaceis squamæformibus, sepalis lineari-lanceolatis lateralibus in mentum breve obtusum productis, petalis oblongis incurvis pluriès latioribus, labello obovato subrotundo plano retuso apiculato integerrimo.

═══════════════════

SOME of the finest species of this genus are found in the division which we formerly characterised (Plate 27) under the name of EUDENDROBIUM, consisting of plants with leafy stems and flowers growing in pairs, or perhaps threes, from the sides of the stem. The division is separated from DESMOTRICHUM, on the one hand, by the lip not being broken up into a tuft of hairs, and, on the other, from STACHYOBIUM, by the flowers not forming racemes.

Of the Eudendrobes, about a third have the lip divided distinctly into three lobes, and these consist for the most part of small-flowered species of little interest, although they also include such charming plants as *D. Ruckeri, sanguinolentum, villosulum,* and *Jerdonianum,* if the two last species are really distinct from each other, as would seem from no mention being made by Dr. Wight of hair upon the stems of the last.

The remainder consist of about 30 known species, in which the lip has no lateral lobes, but forms, when flattened, a circular or oblong plate. Of this division there are three natural groups, of which it is not easy to define the limits, but which the cultivator will readily appreciate. Of the first, the *D. macrophyllum* and *nobile* may be taken as types; of the second, *D. Pierardi* and *pulchellum;* while the third includes the yellow-flowered species, such as *D. chrysanthum* and *aureum.* To the last group belongs the very remarkable plant now figured.

D. albosanguineum, so named from its broad whitish flowers marked in the middle with a rich sanguine stain, is a stout erect plant with knobby stems, rather thicker at the upper than the lower end. It has broad firm leaves (not seen by us in a full-grown state), and from the sides of the stem it produces in pairs very large flowers, of a waxy appearance and consistence, with none of the transparency that belongs to *Pierardi* and its allies. When spread flat, these are full 4 inches in diameter. The sepals are very narrow, and curve inwards, as do the broad banner-like petals, which form a kind of vault over the lip and column. The lip is nearly flat, by no means cucullate, except just at the very base, where it presses against the column. We have had no opportunity of examining the plant any further, and now subjoin Messrs. Veitch's account of it:—

" This species was found by Mr. T. Lobb in open forests on hills near the Atran river, in Moulmein. The description he sent us of it was as follows:—Stems round, jointed, erect; spikes two and three terminal, erect, five and six-flowered; flowers 2½ to 3 inches across, white, with two intense purple spots on the lip; petals also stained with purple at the base. We only received it on the 23rd of April, and in June the flowers you saw expanded; the plants were just bursting into flower when he collected them in February. We have it growing both in pots and on bare blocks, in both which situations it is doing well with us. We find it do well under just similar treatment to that we give to *D. formosum.* It is evidently a free grower, and we have no doubt next spring we shall have it bloom very fine; although the specimen sent you had but two flowers, yet from the *old* spikes it is evident it flowers, as Lobb describes, in fives and sixes on a raceme." We found the flowers to grow in pairs, as we have stated. Perhaps two or three pairs may have been taken for one single inflorescence. If it really forms racemes then it will have to be removed to the Stachyobes.

A CATALOGUE

Of the Dendrobes *belonging to the section* Eudendrobium *having an undivided lip, with their synonymes and horticultural merits.*

Group 1.—GRANDIA.

1. D. macrophyllum *Lindley.*—Manilla.—Flowers very large, bright rose-colour, rhubarb-scented.

2. D. anosmum *Lindley.*—Manilla.—Like the last, but scentless.

3. D. moniliforme *Swartz.*—Japan.—Flowers large, showy, rose-colour, not spotted.

4. D. cœrulescens *Lindley.*—E. Indies.—Flowers showy, rose-colour, with a purple-stained lip. (aliàs *D. Wallichii* of gardens.)

5. D. nobile *Lindley.*—China.—Flowers large, rose-colour, with a purple-stained lip, larger than in the last.

6. D. tortile *Lindley.*—Java ?—Flowers very handsome, violet, with a primrose-coloured lip.

Group 2.—TRANSPARENTIA.

7. D. pulchellum *Roxburgh.*—Sylhet.—Sepals whitish ; petals pink ; most beautiful.

8. D. Devonianum *Paxton.*—Khasija Hills.—Like the last, but much handsomer.

9. D. Pierardi *Roxburgh.*—E. Indies.—Flowers delicate pink ; very pretty.

10. D. cretaceum *Lindley.*—Moulmein.—Flowers chalky-white, with crimson-pencilled lip.

11. D. cucullatum *R. Brown.*—E. Indies.—Very like *D. Pierardi.*

12. D. Egertoniæ *Lindley.*—E. Indies.—Flowers pale pink, very sweet-scented.

13. D. mesochlorum *Lindley.*—E. Indies.—Flowers white, with the petals, &c., tipped with pink, rather sweet-scented.

14. D. crepidatum *Lindley.*—Indian Archipelago.—Flowers white, tipped with pink ; a yellow stain on the lip ; very pretty.

15. D. transparens *Wallich.*—E. Indies.—Flowers pink, transparent, beautiful.

16. D. amœnum *Wallich.*—Nepal.—Flowers delicately white, exquisitely fragrant.
    (aliàs *Limodorum aphyllum* Roxburgh.)

17. D. macrostachyum *Lindley.*—Ceylon.—Flowers rather small, greenish, not handsome.

18. D. gemellum *Lindley.*—Indian Archipelago.—Flowers small, pale yellowish-green ; of no interest. (aliàs *Pedilonum biflorum* Blume.)

19? D. foliosum *Brongniart.*—Java.—I have seen in Reinwardt's Herbarium fragments of what seems to be this plant, but am unable to determine whether or not it is a Dendrobium. There is a small, reflexed, tongue-like appendage on the lip, which excites suspicion that it may be an axillary-flowered Appendicula. It should be compared with *D. auriferum*, a curious Chinese plant, of which I have never seen a specimen.

20. D. candidum *Wallich.*—Khasija Hills.—Flowers small, pure white, sweet-scented.

21. D. nutans *Lindley.*—Ceylon.—Flowers small, white or greenish, with a yellow lip ; stem hairy.

22. D. stuposum *Lindley.*—E. Indies.—Flowers small, white, with a deep orange callus below the point of the lip.

23. D. connatum *Lindley.*—Java.—Flowers whitish green.
    (aliàs *Onychium connatum* Blume.)

Group 3.—CHRYSANTHA.

24. D. chrysanthum *Wallich.*—Nepal.—Flowers deep yellow, with a double purple blotch on the lip.

25. D. Paxtoni *Lindley.*—Khasija Hills.—Flowers orange-yellow, with a deep brown spot in the middle of the lip.

26. D. ochreatum *Lindley.*—Khasija Hills.—Flowers rich yellow, resembling the last.
    (aliàs *D. Cambridgeanum* Paxton.)

27. D. albo-sanguineum *Lindley*.—Moulmein.—Flowers very large, cream-coloured, with two deep crimson stains on the flat lip.

28. D. aureum *Lindley*.—E. Indies.—Flowers pale yellow or white, very fragrant.
(aliàs *D. heterocarpum* Wallich.)

29. D. rugosum *Lindley*.—Java.—Flowers pale yellow.
(aliàs *Grastidium rugosum* Blume.)

30. D. salaccense *Lindley*.—Java.—Flowers deep yellow.
(aliàs *Grastidium salaccense* Blume.)

GLEANINGS AND ORIGINAL MEMORANDA.

362. HELCIA SANGUINOLENTA. *Lindley.* A Peruvian epiphyte of the Orchidaceous Order, with greenish flowers banded with brown, and a white lip marked with broken crimson veins. Introduced by the Horticultural Society. (Fig. 182; *a*, lip magnified; *b*, column do.; *c*, pollen-masses.)

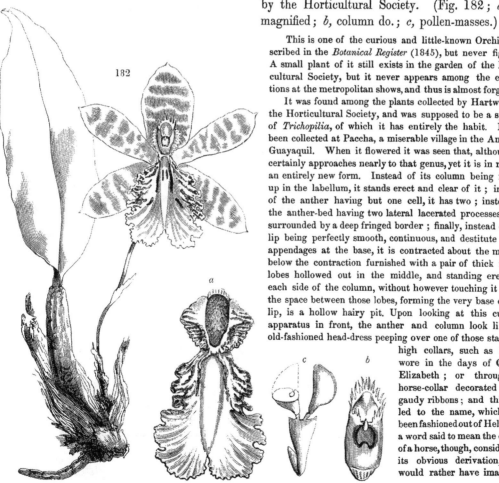

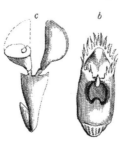

This is one of the curious and little-known Orchids described in the *Botanical Register* (1845), but never figured. A small plant of it still exists in the garden of the Horticultural Society, but it never appears among the exhibitions at the metropolitan shows, and thus is almost forgotten.

It was found among the plants collected by Hartweg for the Horticultural Society, and was supposed to be a species of *Trichopilia*, of which it has entirely the habit. It had been collected at Paccha, a miserable village in the Andes of Guayaquil. When it flowered it was seen that, although it certainly approaches nearly to that genus, yet it is in reality an entirely new form. Instead of its column being rolled up in the labellum, it stands erect and clear of it; instead of the anther having but one cell, it has two; instead of the anther-bed having two lateral lacerated processes, it is surrounded by a deep fringed border; finally, instead of the lip being perfectly smooth, continuous, and destitute of all appendages at the base, it is contracted about the middle, below the contraction furnished with a pair of thick fleshy lobes hollowed out in the middle, and standing erect on each side of the column, without however touching it; and the space between those lobes, forming the very base of the lip, is a hollow hairy pit. Upon looking at this curious apparatus in front, the anther and column look like an old-fashioned head-dress peeping over one of those starched high collars, such as ladies wore in the days of Queen Elizabeth; or through a horse-collar decorated with gaudy ribbons; and this has led to the name, which has been fashioned out of Helcium, a word said to mean the collar of a horse, though, considering its obvious derivation, one would rather have imagined it to signify his traces. The relationship of the plant is evidently greatest to *Trichopilia*; it is, however, also an associate of *Aspasia*, from which it differs in not having the lip united to the column, and in its deep-fringed anther-bed.

363. SPIRÆA LAXIFLORA.   *Lindley.*   A very pretty shrub from Nepal, with white flowers appearing in July. Belongs to Roseworts. Introduced by the Horticultural Society. (Fig. 183.)

183

This very distinct shrub was first described in the *Botanical Register* for 1839. It was there stated to resemble *S. vacciniifolia* in the form of the leaves, and the colour of their underside, but they are long-stalked and rather glaucous above, while the flowers are arranged in large, loose, straggling panicles; the petals are moreover reflexed. The species differs from *S. fastigiata* of Wallich, in the leaves having much longer stalks, being more ovate, with crenatures rather than taper-pointed serratures, and in the panicles being far more lax. It is perfectly hardy, and being much more dwarf than most of the shrubby Spiræas, is well adapted for the front of shrubberies or for decorating parterres of a mixed nature.

364. FORTUNÆA CHINENSIS.   *Lindley.*   (*aliàs* Platy-carya strobilacea *Zuccarini.*)   A greenhouse shrub with pinnated leaves and cones of green flowers. Belongs to Juglands. Native of China, upon the hills of Chusan and Ningpo. "The Chinese use the fruit of this to dye the black colour of their clothes."

An empty cone of this singular plant was received some years ago from Dr. Cantor, by favour of Lord Auckland, then Governor-General of India; and it was at that time supposed to belong to some unknown Conifer. Mr. Fortune rediscovered it, and sent home good seeds and dried specimens, and it now proves to be a plant like a *Rhus* in aspect, but in reality a most curious genus of the natural order of Juglands. If, indeed, we could suppose a walnut to be pressed flat, reduced to the size and texture of a seed of the Alder tree, and then many such to be collected into a small cone, composed of hard, brittle, sharp-pointed scales, we should form artificially what nature has produced in this plant. The annexed figure will explain more particularly these facts, if it is borne in mind that Fig. 1 is a cone; 2, one of the ripe nuts taken out and much magnified; and 3, an inside view of the same; for it will be obvious that the latter might almost be taken for a walnut viewed through a diminishing glass. This shrub or tree, for it is uncertain which it is, is perfectly distinct from all the other genera of Juglands in having its male flowers in catkins, like those of a willow, composed of narrow scales, hairy, and apparently white inside, with four small stamens at their base (Fig. 5). The young nuts are small lenticular bodies with a wing on each side, a minute superior four-toothed calyx, and a pair of short-spreading stigmas (Fig. 4); as the most remarkable genus found by Mr. Fortune during his Chinese expedition, it is proposed to give it the name of its indefatigable discoverer.—*Journal of Hort. Soc.*, vol. i. [At the time this was written the genus Platycarya was unknown to English botanists.]

365. ATROPA ACUMINATA.   *Royle.*   A green-flowered, hardy, herbaceous plant, belonging to the order of Nightshades.   Native of Chinese Tartary at an elevation of 12,000 feet.

This plant is very much like our European A. Belladonna; but its leaves are firmer, narrower, and very much tapered to the point; and the flowers are a pale dull yellow, without a trace of the chocolate colour so characteristic of the European Belladonna. The berries are not distinguishable. It is a hardy perennial, growing freely in any common garden soil, and easily increased either by seeds, or by dividing the old roots when in a dormant state. It grows about 4 feet in height, and flowers in June and July. It is only valuable as a distinct kind of Deadly Nightshade, with yellow flowers.—*Journal of Hort. Soc.*, vol. i.

366. CALYCANTHUS OCCIDENTALIS.   *Hooker.*   A hardy deciduous shrub, with brown, slightly scented flowers.   Native of California.   Introduced by the Horticultural Society.   (Fig. 184.)

Raised from seed sent home by Hartweg, from California, under the name of *Calycanthus macrophyllus*, and said to be a shrub six feet high, growing along rivulets near Sonoma, California. A pale green bush; leaves oblong, acuminate, smooth, and coloured alike on both sides, with short stalks; obtuse or slightly cordate at the base, slightly scabrous above. Flowers solitary, brownish red, larger than usual, with a subacid unpleasant odour. Bracts numerous, subulate, revolute, green. Sepals and petals linear-lanceolate, obtuse, the outer spreading or even rolled back, the inner erect, few, of unequal lengths, incurved, completely concealing the stamens. This species is rather tender, with a handsomer foliage than other "Carolina Allspices," but without their delicious fragrance. It is more an object of botanical than horticultural interest. It flowers in June and July.—*Journal of Hort. Soc.*, vol. vi.

184

367. STIGMAPHYLLON MUCRONATUM. *Adrien de Jussieu.* (*aliàs* Banisteria mucronata *De Candolle.*) A yellow-flowered hothouse climber, belonging to the Malpighiads. Native of Mexico.

This is a twining plant with fleshy roots and opposite ovate oblong leaves terminated by a small point. They are of a bright light-green colour, and have a pair of glands on the stalk just where the leaf sets on. The flowers are of a rich canary-yellow, rather larger than a shilling, with spoon-shaped brown petals, delicately fringed and wrinkled; they grow in small clusters. In this country it must be treated as a greenhouse plant. It will succeed best if planted out in the border of the house and trained up the rafters. When kept in a pot it is necessary to have a trell made, round which the branches can be trained. Any good garden soil seems to suit it, and it strikes readily enough from cuttings. As it has a thick fleshy root, it requires but a small supply of water after it has made its growth for the season. If the species flowers freely it will be a desirable plant owing to its neat habit, which is that of the *Stigmaphyllons* in cultivation. —*Journal of Hort. Soc.*, vol. i.

368. CLEISOSTOMA BICOLOR. A straggling epiphytal Orchid from Manilla, with pink and

purple flowers.   Introduced by Messrs. Veitch.   Flowered at Chatsworth in July.   (Fig. 185 ; *a*, a raceme of the natural size ; *b*, one of the flowers magnified.)

C. bicolor, scapo elongato simplici apice tantum florido, sepalis lateralibus abbreviatis falcatis planis apice rotundatis, labelli lobis lateralibus truncatis angulatis intermedio parvo ovato decurvo, calcare crasso brevi scrotiformi saccato, appendice membranaceâ triangulari bidentatâ.

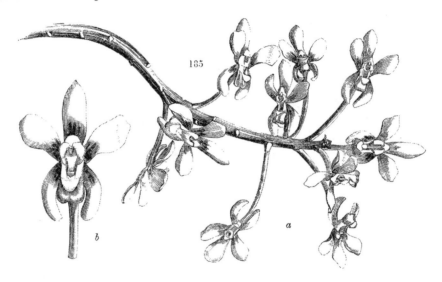

Of the many species of Cleisostom now known there is hardly one worth cultivation ; and this forms no exception to the rule.   We have not seen the stem or leaves, but the flowers are dirty pink, small, and collected in a short raceme at the end of a peduncle eighteen inches long without a branch.   Of the sepals the upper is linear, oblong and straight ; the lateral are more blunt, falcate, and rounded at the end ; the petals are obovate ; all these parts have a dingy, deep purple stain at the base.   The lip is shorter than the sepals ; the lateral lobes are erect, truncate and sharply angular, the middle one is smaller, ovate and recurved ; the bag is very blunt, double, and more fleshy than the sides ; inside, from the back of the bag, rises a triangular, membranous, bifid appendage.   The plant was bought for Chatsworth at a sale of some of Messrs. Veitch's plants in September the 29th, 1848.

369. EUCALYPTUS COCCIFERA. *J. Hooker*.   A hardy glaucous Van Diemen's Land tree, with white flowers.   Belongs to the Myrtleblooms (*Myrtaceæ*).   Introduced about 1842, by Ronald Gunn, Esq.

This plant was exhibited in flower at the June meeting of the Society by Messrs. Veitch, under the name of *Eucalyptus montana*.   It has lived for many years in the garden against a south wall without being injured, but the plants in the open borders dwindled away and died.   According to Messrs. Veitch it is perfectly hardy at Exeter, where it already forms a fine open spreading tree, twenty feet high, and from fifteen to eighteen feet through.   It has grown there for eleven years, and when in flower in June looks like an apple-tree or pear-tree loaded with blossoms.   According to Dr. Hooker it is a species inhabiting the highest mountains of Van Diemen's Land, where it becomes a bush, or small tree, about ten feet high.   It is both Nos. 411 and 1076 of Mr. Gunn's collections, and appears to be sometimes glaucous, sometimes green.   In the garden it has a thick bluish bloom spread over every part.   The branches are purplish brown and slightly rugged.   The leaves oblong, more or less narrow, long-stalked, usually equal-sided, and are most commonly extended at the point into a long and slender awn, by which it is readily recognised.   The flowers are produced on short compressed peduncles in clusters of three to five; the tube of the calyx is pear-shaped, and the lid rugged and convex, but slightly concave in the centre.   The fruit when ripe is nearly hemispherical, with a slightly-raised even border.   As far as can be at present ascertained this may be expected to prove one of the hardiest of the Van Diemen's Island trees.—*Journal of Hort. Soc.*, vol. vi.

370. LYSIMACHIA CANDIDA.   *Lindley*.   A hardy herbaceous plant with white flowers, belonging to the order of Primworts.   Raised from the soil contained in a box sent from China.

This is a dwarf, compact, dark-green herbaceous plant, growing about a foot high.   It is perfectly smooth.   The radical leaves are narrowly oval, tapering into the stalk, and about four inches long; those of the branches are very narrow, and somewhat spathulate; all of them are very obscurely toothed at the edge, or show some tendency to be so, and are marked by scattered dark-purple dots, which are not seen unless the leaves are viewed by transmitted light. The flowers grow in close racemes, are white, and have much the appearance of those of *L. ephemerum*, but the corollas are much larger. From the short time it has been in the garden it is impossible to state what its proper mode of treatment may be. It will in all probability prove hardy, or at least enough so for bedding out in the flower-garden. It appears to be a plant of free growth, and likely to succeed in any sort of soil. From the profuse manner in which it blossoms, it will doubtless be abundantly multiplied from seed.—*Journal of Hort. Soc.*, vol. i.

371. ACACIA BOMBYCINA. *Bentham.* A fine silky-leaved New Holland shrub, from Swan River, of the Leguminous Order. Flowers bright yellow. Raised from seeds received from Mr. Drummond. (Fig. 186.)

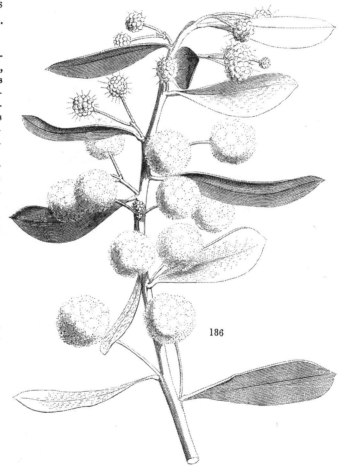

186

A. bombycina; (§ Phyllodineæ, Uninerviæ, Latifoliæ) pube minutâ sericeâ, ramulis subangulatis, phyllodiis obovatis oblongisve subobliquis obtusissimis marginulatis eglandulosis v. obscurè uniglandulosis uninervibus penniveniis, capitulis solitariis v. paucis breviter racemosis multifloris. Phyllodia 1½, 2¼, pollicaria.— *Bentham in litt.*

This handsome species has been for some time in Gardens under the false name of *A. podalyriæfolia.* It forms a small bush, densely covered with a grey silky hairiness. The phyllodes are obovate, tapering to the base, very soft, one-nerved, and usually with a gland a little below the middle of the upper edge. The rich yellow balls of flowers are placed on stalks shorter than the phyllodes, and are either single or in clusters of from 2 to 4 or even 5. Mr. Bentham, who has been good enough to examine the plant, remarks, " It belongs to the *Uninerviæ,* and is near *brachybotrya* and *podalyriæfolia,* differing from the former in its much larger leaves and the silkiness of the pubescence, from the latter in its heads either solitary or few in a short raceme; but positive characters cannot well be given without specimens in flower, as the calyx and corolla often give very good distinctive marks."

Unfortunately our flowering specimens have been mislaid; but there can be no doubt of the distinctness of the species from all as yet in books.

372. AERIDES FLAVIDUM. A handsome fragrant Orchidaceous epiphyte, with yellow and pink flowers. Native country unknown. Flowered with A. Kenrick, Esq.

A. flavidum (A. quinquevulnera facie); labelli cornuti laciniis lateralibus rotundatis integerrimis intermediâ breviore bifidâ glabrâ.

We have received of this three flowers only, with a statement that the plant much resembles *A. quinquevulnera.* They are glutinous and very fragrant; the lip is quite different from that of any species with which we are acquainted, the lateral lobes being rounded and entire, while the middle lobe is much shorter and two-lobed. Of the lip the horn is green the lobes pale yellow; the petals and sepals are white dashed with pink.

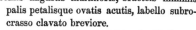

373. ANGRÆCUM MONODON. A small pink-flowered African epiphyte, with distichous leaves. Flowered with M. Pescatore of Paris. (Fig. 187, much magnified.)

A. monodon; subacaule, foliis distichis oblongis obliquè bilobis, racemis angustis multifloris, bracteis minimis membranaceis rotundatis cucullatis, secundo basi dente solitario aucto calcare palis petalisque ovatis acutis, labello subrocrasso clavato breviore.

This curious little species was obtore, from whom we received a specimen oblong, distichous leaves, and small seeds of a horse-chesnut. Its specific tooth which stands on the lip in front of the accompanying cut, which represents tained from Gabon in Africa, by M. Pesca a year and a half ago. It has roundish, reddish flowers, smelling something like the name refers to the presence of a small erect the opening into the spur, as is shown in a flower six times larger than natural.

187

374. ATACCIA CRISTATA. *Kunth.* Tacca Rafflesiana *Jack.*) A dingy-Malacca. Belongs to the Natuduced by the Royal Botanic (*aliàs* Tacca cristata *Jack; aliàs* flowered tuberous stove-plant from ral Order of Taccads. Intro-Garden, Kew.

Both Endlicher and Kunth, though *Ataccia* for the entire-leaved species of propriety of the separation. I am in-they follow Presl in adopting this genus *Tacca,* yet express their doubts as to the competent to pronounce, through a want of recent specimens of the original *Tacca,* on the value of the distinctions: but, judging from the figures and dried specimens, the difference is more in habit than in essential character. *Tacca* has multifid leaves and tuberous roots, and may be considered an annual plant. *Eutaccia* has entire leaves, a short subterraneous conical stem or caudex, quite different from the tubers of the former. There is no difficulty, therefore, in recognising the respective genera.

A. cristata, the subject here figured, has been long cultivated in the stove of the Royal Gardens of Kew, under the name of *Tacca integrifolia* Gawl., and is a native of the Malay Islands and Archipelago. *Tacca aspera* Roxb. (*T. integrifolia* Gawl. in *Bot. Mag.* t. 1488, and of Roxb. Coromandel plants, vol. iii. t. 257), from Chittagong, may be known by the short scape or flower-stalk, which, as well as the petioles, are scabrous. *Tacca lævis* Roxb., from Silhet Gualpara, and Chappedong (Wall.) and Assam, is easily recognised by the four sessile uniform leaves of the involucre, and small and slender habit. *Tacca lanceæfolia* Zoll. (*Ataccia* Kth.), is probably a variety of the latter.—All these are Indian: but I possess another and distinct species from Demerara, South America, with a creeping rhizoma! There are few more remarkable-looking plants in cultivation than our *Ataccia cristata.* Root a few coarse fibres, issuing from a short, underground, conical, descending caudex or rhizoma, marked with the rings or scars of fallen leaves, and here and there throwing out small tubers or gemmæ. Leaves three or four, all from this short caudex. Petioles semiterete, smooth: the blade oblong, acuminate, dark purple-green, penninerved, nerves mostly prominent beneath. Scape about as long as the leaves, erect, stout, angled, dark purple, smooth: terminated by a large, dark purple, four-leaved, membranaceous involucre: the two outer leaflets opposite, sessile, ovato-acuminate, striated, patent, two inner placed side by side, erect, very large, greenish, striated, reticulated, edged with purple; the shape broadly ovate, acute, but tapering into a long, narrow, deep purple base. Peduncles numerous, dark purple, about two inches long, terminated each by a single flower and forming a drooping unilateral umbel: these floral peduncles are accompanied by several (external) long, tapering, filiform sterile ones, six inches long, which spread out in their lower portion, while the rest of the tendril-like peduncle droops. Perianth dark purple: the tube turbinate, six-angled, for the greater part united with the ovary; the limb sexpartite, suddenly reflexed; the segments or lobes in two series, outer smaller, the inner larger, all ovato-rotundate, acute, striated, the rim of the mouth forming a crenated ring. Stamens six, within the mouth of the tube: filament broad, the margin lamellate and plaited, the back cohering with the perianth; anther cucullate, two-celled: pollen globose. Ovary adherent with the calyx tube, one-celled, having three longitudinal, furrowed, parietal placentæ, bearing several ovules. Style short, conical, six-furrowed. Stigma of three, broad obcordate, green, reflexed, plaited lobes; the edges of the plaits ciliated. This singular tropical plant is of easy cultivation. It grows and flowers freely in a moist, warm stove. A mixture of light loam and peat-soil suits it, and, being a native of moist places, it requires a copious supply of water. It increases freely by offsets, which are produced from the sides of the erect rhizome-like caudex; these offsets, when separated, root readily in small pots placed in a close moist atmosphere.—*Bot. Mag.,* t. 4589.

375. PHILADELPHUS SATSUMI. *Siebold.* A hardy deciduous shrub, with white flowers. Native of Japan. Belongs to the Order of Syringas. Blossoms in July. (Fig. 188.)

We have failed to discover in what work this plant has received the name by which it has been sent to this country. It is nearly allied to *Ph. laxus;* but seems to be distinct from that and all other American species. It forms a very graceful bush, with a good foliage of a dark green colour, with the upper leaves very long, narrow and undivided. The foliage is slightly hairy on the underside, the lower leaves oval-lanceolate, acuminate, with a few shallow very acute serratures.

The flowers grow singly or in pairs at the end of weak, slender shoots ; but, if a Japanese specimen, without a name, given us by the late Professor Zuccarini, should be the same as this, they will appear hereafter in long interrupted racemes with linear or almost filiform bracts. The calyx is smooth, with divisions very variable in length. The styles are divided almost to the base. As a hardy deciduous shrub, this must be regarded as an acquisition.

376. ZAMIA LINDLEYI. *Warczewicz.* A hothouse shrub, with pinnated narrow leaves, from Veragua. Belongs to Cycads. Introduced by Mr. Warczewicz.

This species has a somewhat cylindrical stem, from six to seven feet high, equally pinnated leaves, consisting of many pairs of linear, sharp-pointed, acuminate, entire leaflets, and a hispid

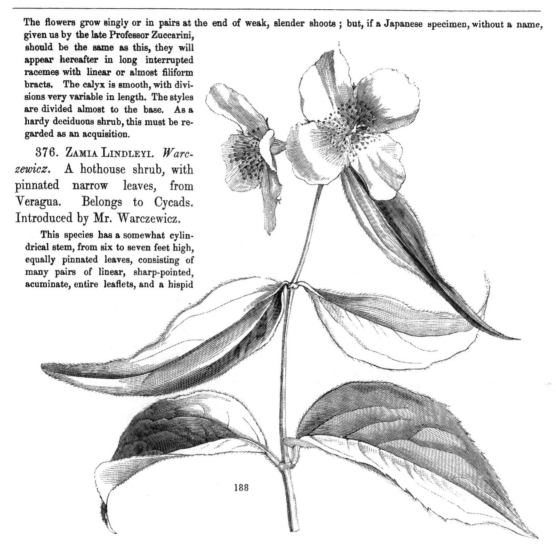

188

petiole. Found with the next on the Cordillera of Veragua, at the elevation of from 5000 to 7000 feet above the sea. —*Allgem. Gartenzeit.*, May 10th, 1851.

377. ZAMIA SKINNERI. *Warczewicz.* A hothouse shrub with pinnated broad leaves, from Veragua. Belongs to Cycads. Introduced by Mr. Warczewicz.

The same traveller found this growing in company with *Zamia Lindleyi.* The stem is from four to six feet high, broader at the bottom than the top. The leaves are equally pinnated, and consist of many pairs of elliptical-lanceolate leaflets, acute at each end and serrated near the point. Their petiole is prickly.—*Allgem. Gartenzeit.*, May 10th, 1851.

378. CEANOTHUS CUNEATUS. *Nuttall.* A half-hardy evergreen shrub, from California. Flowers white. Belongs to the Order of Buckthorns.

Raised from seeds received from Hartweg in June, 1848, marked Ceanothus sp., with white flowers, a shrub six or eight feet high, from the Sacramento mountains. It is tender, and will not live in the open border. It flowers in May. This shrub is described as follows by Mr. Nuttall :—" A shrub six to ten feet high, with somewhat thorny greyish terete branches, very closely interwoven, sometimes forming thickets. Leaves half an inch or more in length, and about two lines wide; very rarely with one or two teeth near the extremity; the numerous regular, simple, and oblique veins rather conspicuous on the lower surface. Flowers in small axillary umbels: the peduncles and pedicels increasing in length as the fruit ripens. Calyx and corolla white: petals cucullate, unguiculate. Styles united above the middle, and then spreading. Fruit as large as an ordinary pea, sub-globose; the exocarp somewhat pulpy, with three rather soft horn-like projections from the summit of the angles: the coherent base of the calyx unusually large. Seeds even on both sides, black, polished.   The whole plant (like several succeeding species) exhales a balsamic odour, and the mature fruit is covered with a bitter varnish.''

It is said to grow as far north as " the dry gravelly islands and bars of the Wahlamut river above the falls," in Oregon; but it is best known from more southern regions, Hartweg's discovery of it in California having been anticipated by the naturalists with Captain Beechy, and by Dr. Coulter, of whose dried plants it is No. 110. In our gardens it betrays a tender climate, for it is much more impatient of cold than the other Californian species, than which it is far less attractive, its scanty white flowers producing a shabby appearance, for which the leaves and scrubby aspect of the species do not compensate.—*Journal of Horticultural Society*, vol. vi.

379. DENDROBIUM CLAVATUM.   *Wallich*, Cat. No. 2004.   A magnificent epiphyte with bright yellow flowers and a dark eye.   Native of Assam.   Introduced by Thomas Denne, Esq.   (Fig. 189, a single flower forced open and magnified.)

D. clavatum (Stachyobium) ; caulibus teretibus pendulis, foliis . . . ., racemis lateralibus laxis 5-floris flexuosis basi squamatis, bracteis membranaceis oblongis cucullatis internodiis æqualibus, sepalis lineari-oblongis, petalis obovato-oblongis rotundatis subundulatis, labello transverso leviter trilobo pubescente margine recto ciliato.

This very fine plant was received from Assam in February last, by Thomas Denne, Esq., of Hythe in Kent, and flowered with him in May. The stems are terete, from eighteen inches to two feet long ; the leaves we have not seen. The flowers appear in fives, in close heads, from among some hard scales ; and are separated by large membranous bracts almost as in *D. densiflorum ;* when the racemes are full grown their rachis is zigzag, and the broad membranous bracts are full as long as the joints of the rachis. The expanded flowers are about two inches across when flattened, but as the parts spread but little from the column they appear smaller ; they are of a rich orange-yellow, with a broad double brown blotch in the middle of the lip. The sepals are much narrower than the petals, which are not at all fringed. The lip, when flattened, is broader than long, slightly three-lobed, round, hairy over all the upper surface, and strongly ciliated, though not fringed, at the edge. Mr. Denne most truly says, that " It is certainly the handsomest of the orange Dendrobes, being superior to *D. Paxtoni* in size and texture and also in the markings of the lip, though it has not the fimbriated edge." The affinity of this species is with *D. fimbriatum* and *moschatum*, to the latter of which we were formerly led by bad specimens to refer it as a synonyme. From *D. fimbriatum* it differs in having large membranous bracts, and no deep fringes to the lip. In its bracts it agrees with *D. moschatum*, and in the flowers appearing from within hard scales, but the lip has not the inflexed edge and slipper-like form of that species, and the racemes are much shorter.

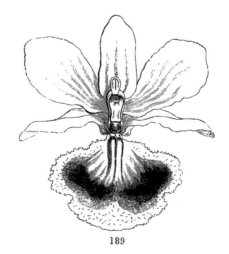

189

PLATE 58.

L. Constans del. & Zinc.

Printed by C.F. Cheffins, London.

[PLATE 58.]

THE LODDIGES LILY.

(LILIUM LODDIGESIANUM.)

————•————

A handsome Hardy Bulbous Plant from the CAUCASUS, *belonging to the* LILIACEOUS *Order.*

═══════════════

Specific Character.

THE LODDIGES LILY. Leaves close, alternate, spreading, here and there whorled, ovate-lanceolate, rather obtuse, on the underside, especially at the edge and veins slightly downy, the uppermost gradually smaller. Raceme erect, few-flowered. Flowers drooping, two or three times as long as their stalk. Divisions of the flower rolled back.

LILIUM *LODDIGESIANUM;* foliis confertè sparsis patentibus hinc inde subverticillatis ovato-lanceolatis obtusiusculis subtus præsertim in margine venisque puberulis supernè gradatim decrescentibus, racemo erecto paucifloro, floribus cernuis pedicello duplo triplove longioribus, calycibus revolutis.—*Kunth, enum.* 4. 261.

————————

L. Loddigesianum : *Römer and Schultes, Systema,* 7. 416. *Morren, Annales de Gand,* vol. ii., p. 363, t. 85.

════════════

THIS fine hardy bulbous plant was received by the Horticultural Society on April 3, 1842, from Mr. de Hartwiss, of the Imperial Gardens, Nikita, in the Crimea, under the name of *Lilium monadelphum.* A few months later it came from Dr. Fischer, of St. Petersburg, under the same name. Yet it is in no degree monadelphous; on the contrary, its stamens are distinct to the very base.

Lilium monadelphum was so called by Bieberstein in his account of the Caucasian flora, and described as a plant the size of *Lilium album,* with flowers of the same size and form, but yellow, and with the filaments united sometimes into a tube as long as the ovary, sometimes into a mere ring. Römer and Schultes add that cultivated plants raised from Crimean seeds grew from 2 to 4 feet high, with campanulate flowers, tubular at the base, and spreading at the point, but in no degree rolled back; the petals were quite yellow with no spots, and the stamens were joined into a tube rather longer than the ovary. It would therefore seem clear that our plant cannot be *L. monadelphum.*

It was, however, figured under that name by the late Mr. Ker, at t. 1405 of the Botanical Magazine, and Bieberstein afterwards praised the representation as a good one. But Römer and Schultes, unable to reconcile with that author's account a plant in which the divisions of the flowers are revolute like a Turk's cap, and spotted, while the filaments are wholly disunited, proposed to call the latter, now before us, *L. Loddigesianum,* because Mr. Loddiges had first raised it from Russian seeds. In this Prof. Kunth acquiesced.

Nevertheless the Russian Botanists Fischer, Meyer, and Avé Lallement have united *L. Loddigesianum* and *L. monadelphum,* describing their plant as $3\frac{1}{2}$—$5\frac{1}{2}$ feet high, with from 1—27 flowers, and stamens united at the base, all which is at variance with our plant; at the same time they created a *L. Szovitzianum,* from Colchis, very near *L. monadelphum,* with free stamens, and flowers like wax in colour and texture. Thus far it corresponds with the plant now before us; but the above authors add that the flowers are spotted inside with dark purple, the style twice as long as the ovary, and the leaves scabrous at the edge, in which respects this disagrees. Upon the whole, therefore, we leave the name *L. Loddigesianum* as we find it, till some one shall succeed in settling the intricate synonymy of this genus, when it is probable that a great reduction of so-called species will take place.

In the meanwhile we venture to ask what difference there is between *L. Loddigesianum* and *L. pyrenaicum ?* beyond size and the spotting of the flowers.

PLATE 59.

L.Constans del. & Zinc.

Printed by C.F.Cheffins, London.

[PLATE 59.]

THE ARIZA PLANT.

(BROWNÆA ARIZA.)

———◆———

A Superb Hothouse Tree from CENTRAL AMERICA, *belonging to the* LEGUMINOUS *Order.*

═══════════════════

Specific Character.

THE ARIZA PLANT.—Leaves in 6 or 8 pairs, oblong-lanceolate, with long points, usually narrowed at the base, the shorter of the lower couples cordate at the base. Bractlets connate, downy outside, 3 times as long as the tube of the calyx. Stamens 11, not so long as the corolla, free from their very base.

BROWNÆA *ARIZA ;* foliis 6-8-jugis oblongo-lanceolatis longè cuspidatis basi plerisque angustatis jugorum inferiorum brevioribus basi cordatis, floribus densè capitato-spicatis, bracteolis connatis extus tomentosis calycis tubum triplo superantibus, staminibus 11 corollam vix æquantibus à basi liberis.—*G. Bentham.*

═══════════════════

Brownæa Ariza : *Bentham in Plantæ Hartwegianæ*, p. 171, no. 961*.

═══════════════════

ONE of the finest tropical trees in cultivation, and more especially valuable, because it produces its magnificent heads of scarlet flowers without difficulty. The specimen now figured was obtained from the Garden of the Horticultural Society in June last. The collector Hartweg, from whom it came, says that it inhabits woods near Guaduas, in the province of Bogota, at the elevation of 1400 feet above the sea; that the people call it Ariza, and that it forms a tree from 30 to 40 feet high.

It is nearly related to the *Brownæa grandiceps* of the Caraccas, from which Mr. Bentham distinguishes it by its bracts and flowers being larger, the proportions of the floral organs different, and the stamens wholly distinct from each other. To this an inspection of the living plant enables us to add that the leaflets are larger and flatter, with a thicker texture. The claws of the petals are as long or longer than the lobes of the calyx, of which there are four not three. The following remarks, applied in the Botanical Register to *Brownæa grandiceps*, are equally suited to the Ariza :—

All the species of this genus are stove shrubs, inhabiting the hottest parts of America. Their flowers are produced in a short spike, tier above tier, every day witnessing the expansion of a new tier above those of the former days, till at last the whole mass becomes a globe of living and glowing crimson.   This brilliant head appears on the side of the main stem, among the leaves, which at that time present a singular phenomenon.   Every evening they rise up and lift themselves from the blossoms to expose them to the dew, so that each morning these beautiful objects lie uncovered ; but as day advances the leaves gradually droop, and bend down over the flowers to guard them from the rays of the sun.   Who can imagine the gorgeousness of an equinoctial forest at midnight, with the veils thus lifted off myriads of flowers of every form and hue, all hidden from our gaze by this or other means during the hours of tropical sunlight, whose brilliancy would be death to their tender texture and delicate colours ?

This tree must be grown in the damp stove.   When its seeds are good they are easily raised if sown in light soil, and plunged in a tan-pit or hot-bed.   A rich free soil that will not get hard or sour is the best for its after-growth.   It is only in a large house that it can thrive well for any length of time, and be seen in its greatest beauty.   Then if planted out in the border, or in a large tub with sufficient room for its leaves to develope freely, it forms a magnificent object, not perhaps much inferior to *Amherstia nobilis*.

L. Constans. del. & Zinc.

Printed by, C.F. Cheffins London.

[PLATE 60.]

THE ROSY AIR-PLANT.

(AERIDES ROSEUM.)

———◆———

A Hothouse Epiphyte from the EAST INDIES, *belonging to the Natural Order of* ORCHIDS.

𝔖𝔭𝔢𝔠𝔦𝔣𝔦𝔠 𝔆𝔥𝔞𝔯𝔞𝔠𝔱𝔢𝔯.

THE ROSY AIR-PLANT.—Leaves coriaceous, channelled, distichous, blunt and two-lobed at the point. Spikes dense, recurved. Sepals, as well as the petals, which are longer and narrowed at the base, acute. Lip lozenge-shaped, acuminate, flat, entire, with a short conical incurved spur. Ovary three-winged, as long as the lip.

AERIDES *ROSEUM* ; foliis coriaceis canaliculatis distichis apice obtusis bilobis, spicis densis recurvis, sepalis petalisque longioribus basi angustatis acutis, labello rhomboideo acuminato plano integerrimo, calcare brevi conico incurvo, ovario trialato labelli longitudine.

Aerides roseum : *Loddiges.* A. affine : *Hooker in Botanical Magazine*, t. 4049, not of Wallich.

THIS noble plant occurs not uncommonly in gardens under the erroneous name of *Aerides affine*, and has been figured as such in the *Botanical Magazine*. It is, however, essentially distinct, as will be shown presently. We first saw it, some years since, in the possession of Messrs. Loddiges, with whom a dark variety was marked No. 1530, India. Since that time it has appeared in many collections. The specimen now figured was from Mr. Conrad Loddiges.

Among the more important peculiarities of this plant are the following :—Its leaves are leathery, channelled, and roundly two-lobed. The sepals and petals are acute. The lip is perfectly undivided, and tapers to the point. The triangular, or rather three-winged ovary, is as long as the lip ; and finally the spikes are drooping, or curved below the horizontal line.

On the other hand *Aerides affine*, of which wild specimens from Dr. Wallich, in all respects agreeing with the figure in the *Sertum Orchidaceum*, are now before us, has truncated leaves, the ends of which are even jagged, of which in *A. roseum* there is no sign. The sepals and petals are remarkably blunt—almost rounded. The lip is more or less toothletted, not unfrequently even three-lobed, and partially imitating the bluntness of the sepals. The ovary is much shorter than the lip,—not half its length; and, finally, the spikes are stiff and erect, by no means drooping gracefully. These differences render it impossible to regard the two plants as mere forms of each other.

Another plant closely allied to these is the *Aerides maculosum*, figured in the *Botanical Register* for 1845, t. 58. This differs in having flowers loosely arranged, larger, more spotted, and generally somewhat panicled. The lip has, moreover, at its base two small flat spreading acute lobes; the same lobes occur, no doubt, in *A. affine* and *roseum*, but they are smaller, erect, and rounded.

Finally, Dr. Wright has lately published an *Aerides Lindleyanum*, with short leaves, short erect racemes, flowers far larger than in the allied species, and a very distinctly serrated plaited lip. Of this fine plant, which we have not seen, he gives the following account :—

"Leaves fleshy, coriaceous, sub-elliptic, oblong, oblique, deeply emarginate at the apex; racemes erect, many-flowered; sepals and petals obovato-suborbicular, anterior sepals somewhat larger, and, like the lip, thick and coriaceous; lip three-lobed, attached to the point of the prolonged base of the column; lateral lobes small, ovate, middle one large, ovate, ventricose above, crisp on the margins, with a large fleshy lobe at the base, closing the spur; spur short, rigid, inflexed under the lamina; capsules large, obovate, long pedicelled; flowers pinkish-lilac, deeper on the axis, fining off to nearly white on the margins; lip the same, but much deeper coloured. On the clefts of rocks, bordering the Kartairy Falls, below Kaitie; also on rocky clefts on a high hill over Coonoor, flowering nearly the whole year; at least I gathered it in April, and I have it now (November) in flower in pots in Coimbatore."

In order to put these distinctions in a clearer light, we propose the following short technical characters. The species constitute a well-marked division of the genus Aerides, from which many of those now on record will have to be excluded whenever the genus is revised :—

AERIDES §.   *Labello plano indiviso, nunc basi auriculato.*

1. A. affine *Wallich, Catalogue,* No. 7316; *Lindley, Sertum Orchidaceum,* t. 15; A. multiflorum *Roxb. Fl. ind.,* 3. 475. (?); foliis apice truncatis nunc dentatis, spicis strictis, sepalis petalisque rotundatis, labello rhomboideo sublobato ovario duplo longiore.   (Perhaps not in cultivation.)

2. A. roseum *Loddiges; Paxton, Fl. Garden,* t. 60; A. affine *Hooker in Bot. Mag.,* t. 4049; foliis apice bilobis rotundatis, spicis cernuis, sepalis petalisque acutis, labello rhomboideo integerrimo acuminato ovario trialato æquali.

Var. A. floribus pallidè roseis immaculatis.

Var. B. floribus atroroseis submaculatis.

3. A. maculosum *Lindl. in Bot. Reg.,* 1845, t. 58; foliis apice obliquis obtusis, racemis cernuis subpaniculatis, sepalis petalisque obtusis, abello ovato obtuso plano indiviso basi utrinque unidentato tuberculo indiviso interjecto.

4. A. Lindleyanum *Wight, Figures of Orchidaceous Plants,* t. 1677; foliis brevibus apice obliquis obtusis bilobis, racemis paucifloris strictis, sepalis petalisque carnosis obtusis, labello ovato acuto serrato plicato basi auriculo acuto crasso utrinque dente magno carnoso inflexo interjecto.

GLEANINGS AND ORIGINAL MEMORANDA.

380. SAXE-GOTHÆA CONSPICUA. *Lindley.* An evergreen hardy Coniferous tree of great beauty, from the Andes of Patagonia. Introduced by Messrs. Veitch. (Fig. 190.)

Generic Character. Genus *Coniferarum* monoicum.—Fl. masc. *Antheræ* spicatæ, 2-loculares, apice acuminatæ reflexæ —Fl. fœm. *Strobilus* imbricatus, è squamis acuminatis liberis infra medium monospermis. *Ovulum* inversum, in foveâ squamæ semi-immersum ; *tunicâ primâ* laxâ, ventre fissâ, *secundâ* foramine pervio, *nucleo* apice spongioso protruso. *Galbulus* carnosus, è squamis mucronatis, apice liberis, squarrosis, omninò connatis, plurimis abortientibus. *Semen* nucamentaceum, leviter triangulare, basi tunicæ primæ membranaceæ fissæ reliquiis vestitum. Arbor *sempervirens*, Taxi *facie ;* foliis *linearibus, planis, apiculatis, subtùs lineâ duplici pallidâ notatis.*

This remarkable plant, to which His Royal Highness Prince Albert has been pleased to permit one of his titles to be given,

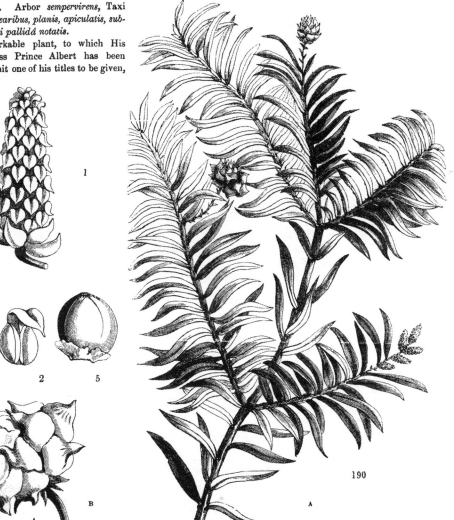

190

and which will probably rank among the most highly valued of our hardy evergreen trees, is a native of the mountains of Patagonia, where it was found by Mr. William Lobb, forming a beautiful tree 30 feet high.   In the nursery of Messrs. Veitch, of Exeter, it has lived in the open air for four years without shelter, and has all the appearance of being well adapted to the climate of England.   The country in which it grows is, indeed, more cold and stormy than any part of Great Britain, as is shown by the following account of it, given by Mr. Lobb in one of his letters to Messrs. Veitch :—

" During my absence I visited a great part of Chiloe, most of the islands in the Archipelago, and the coast of Patagonia for about 140 miles.   I went up the Corcobado, Caylin, Alman, Comau, Reloncavi, and other places on the coast, frequently making excursions from the level of the sea to the line of perpetual snow.   These bays generally run to the base of the central ridge of the Andes, and the rivers take their rise much further back in the interior.   The whole country, from the Andes to the sea, is formed of a succession of ridges of mountains gradually rising from the sea to the central ridge.   The whole is thickly wooded from the base to the snow line.   Ascending the Andes of Comau, I observed from the water to a considerable elevation the forest is composed of a variety of trees, and a sort of cane so thickly matted together that it formed almost an impenetrable jungle.   Further up, amongst the melting snows, vegetation becomes so much stunted in growth, that the trees, seen below 100 feet high and 8 feet in diameter, only attain the height of 6 inches.

" On reaching the summit no vegetation exists—nothing but scattered barren rocks which appear to rise amongst the snow, which is 30 feet in depth, and frozen so hard that on walking over it the foot makes but a slight impression.

" To the east, as far as the eye can command, it appears perfectly level.   To the south, one sees the central ridge of the Andes stretching along for an immense distance, and covered with perpetual snow.   To the west, the whole of the islands, from Guaytecas to the extent of the Archipelago, is evenly and distinctly to be seen.

" A little below this elevation the scenery is also singular and grand.   Rocky precipices stand like perpendicular walls from 200 feet to 300 feet in height, over which roll the waters from the melting snows, which appear to the eye like lines of silver.   Sometimes these waters rush down with such force, that rocks of many tons in weight are precipitated from their lofty stations to the depth of 2000 feet.   In the forest below everything appears calm and tranquil ; scarcely the sound of an animal is heard ; sometimes a few butterflies and beetles meet the eye, but not a house or human being is seen.   On the sandy tracts near the rivers, the lion or puma is frequently to be met with ; but this animal is perfectly harmless if not attacked."

It is from this wild and uninhabited country that many of the fine plants raised by Messrs. Veitch were obtained, and among them the *Saxe-Gothœa, Podocarpus nubigena, Fitz-Roya patagonica,* and *Libocedrus tetragona.*   Of these he writes thus :—

" The two last *(Fitz-Roya* and *Libocedrus)* I never saw below the snow line.   The former inhabits the rocky precipices, and the latter the swampy places between the mountains.   The first grows to an enormous size, particularly about the winter snow line, where I have seen trees upwards of 100 feet high, and more than 8 feet in diameter.   It may be traced from this elevation to the perpetual snows, where it is not more than 4 inches in height.   With these grow the Yews *(Saxe-Gothœa* and *Podocarpus nubigena),* which are beautiful evergreen trees, and, as well as the others, afford excellent timber."

Saxe-Gothæa may be described as a genus with the male flowers of a Podocarp, the females of a Dammar, the fruit of a Juniper, the seed of a Dacrydium, and the habit of a Yew.   Its fleshy fruit, composed of consolidated scales, enclosing nut-like seed, and forming what is technically called a Galbulus, places it near Juniperus, from which it more especially differs in its anthers not being peltate, nor its fruit composed of a single whorl of perfect scales, and in its ovule having two integuments instead of one.   In the last respect it approaches Podocarpus, and especially Dacrydium ; but the exterior integument of the seed is a ragged abortive membrane, enveloping the base only of the seed, instead of a well-defined cup.   In a memorandum in my possession, by Sir William Hooker, I find this distinguished botanist comparing Saxe-Gothæa to a Podocarp with the flowers in a cone—a view which he was probably led to take by the condition of the ovule, and which may be regarded as the most philosophical mode of understanding the nature of this singular genus ; to which Nageia may be said to be a slight approach, and which is not distinguishable by habit from a Podocarp.

In its systematic relations Saxe-Gothæa possesses great interest, forming as it does a direct transition from the one-flowered Taxads to the true imbricated Conifers, without, however, breaking down the boundary between those orders, as I understand them, but rather confirming the propriety of limiting the Coniferous order to those genera which really bear cones instead of single naked seeds.   In the language of some naturalists, Saxe-Gothæa would be called an osculant genus between Taxads and Conifers.

The leaves of this plant have altogether the size and general appearance of the English Yew, *Taxus baccata ;* but they are glaucous underneath, except upon the midrib and two narrow stripes within the edges, which are pale green. The male flowers consist of spikes appearing at the ends of the branches, in a raceme more or less elongated.   These spikes (fig. 1) grow from within a few concave acute scales, which form a kind of involucre at the base.   Each male is a solitary membranous anther, with a lanceolate, acuminate, reflexed appendage, and a pair of parallel cells opening longitudinally.   The female flowers form a small roundish, pedunculated, terminal, scaly imbricated cone.   The scales are fleshy, firm, lanceolate, and contracted at their base, where they unite into a solid centre.   All appear to be fertile, and to bear in a niche in the middle, where the contraction is, a single inverted ovule (fig. 3).   The ovule is globular, with 2 integuments beyond the nucleus ; the outer integument is loose and thin, and wraps round the ovule in such a way that its two edges cannot meet on the under-side of the ovule ; the second integument is firm and fleshy ; the nucleus is flask-shaped, and protrudes a fungous circular expansion through the foramen. The fruit (fig. 4) is formed, by the consolidation of the free scales of the cone, into a solid fleshy mass of a depressed form and very irregular surface, owing to many of the scales being abortive, and crushed by those whose seeds are able to swell ; while the ends of the whole retain their original form somewhat, are free, rather spiny, and constitute so many tough, sharp tubercles. The seed (fig. 5) is a pale brown, shining, ovate, brittle nut, with 2 very slight elevated lines, and a large irregular hilum ; at the base it is invested with a short, thin, ragged membrane, which is the outer integument in its final condition. The nucleus lies half free in the interior, the fungous apex having shrivelled up and disappeared. Since this was written Sir W. Hooker has placed in my hands a sketch of the anatomy of the female flowers of Saxe-Gothæa, by Mr. B. Clarke, who describes the ovule thus :—" Its ovule has the same structure as that of Gnetum, as described by Mr. Griffith, viz. : it has 3 integuments ; the internal protrudes, and forms a sort of stigma, not so obvious as in Gnetum ; the external has constantly a fissure on its posterior, or rather inferior surface, which however does not close as in Gnetum when the ovule advances in growth, nor yet become succulent. Mr. Griffith describes the fissure in the external integument of Gnetum as constantly posterior ; and if the ovules of the strobilus were erect, they would agree with Gnetum in this particular.

Explanation of the Cuts.—*A*, a branch with male and female flowers, natural size ; *B*, various details of the fructification, more or less magnified ; 1, a spike of male flowers ; 2, a male or anther apart ; 3, a scale seen from the inside with the inverted ovule, showing the fungous foramen protruding beyond the primine (outer integument) ; 4, a ripe fruit ; 5, a seed showing the 2 slight elevations upon the surface, and the remains of the ragged primine at the base.—*Journ. of Hort. Soc.*, vol. vi.

381. Spiræa callosa. *Thunberg.* A handsome, hardy, deciduous shrub, with brilliant rose-coloured flowers. Native of the North of China and Japan. Flowers in July and August. Reintroduced by Messrs. Standish and Noble. (Fig. 191.)

In general appearance this resembles the Nepal *Spiræa bella*, but is far more ornamental on account of the brilliant tint of its petals, especially when the flower-buds first begin to expand. The leaves are dark-green, nearly exactly

191

lanceolate, rugose, sharply serrate, tapering to both ends, but entire near the base ; they have a strong tendency to become three-lobed when vigorous ; the serratures are tipped with little brown callosities. On the under side the leaves are glaucous, but not hairy. The flowers are arranged in branched cymes, which usually grow in pairs from the same side of the branch, the lower naked at the base, the upper supported by a long narrow leaf. The calyx is covered with fine silky hairs, and divided into five sharp triangular lobes. The carpels are quite smooth. Mr. Fortune has lately sent this from the northern part of China, but it was long since obtained for the Horticultural Society by Mr. Reeves, one of whose dried specimens is now before us.

382. Hoya Cumingiana. *Decaisne.* A stove scandent shrub, with dense flat leaves and short axillary umbels of greenish-yellow flowers. Native of the Philippines. Blossoms in May and June. Introduced by Messrs. Veitch and Son. (Fig. 192.)

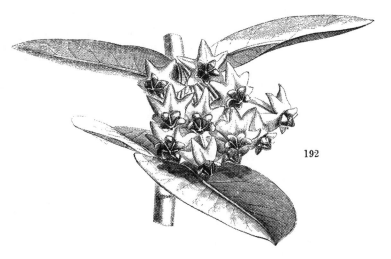

192

At one of the exhibitions in the garden of the Horticultural Society this novelty was produced by Messrs. Veitch and Son. It is an erect bush with closely packed decussating sessile cordate leaves, very slightly downy beneath, and of a somewhat parchment-like consistence. The flowers are destitute of gay colours, the principal tint being yellowish-green, relieved by a coronet of rich purplish brown. It is very distinct from any of the other species in cultivation, and before flowering would not be taken for a Hoya at all.

383. Cathcartia villosa. *J. D. Hooker.* A beautiful annual (?) from Sikkim-Himalaya, with large yellow flowers. Belongs to Poppyworts. Introduced at Kew.

The following is Dr. Hooker's character of this new genus :—Calyx diphyllus, foliolis æstivatione imbricatis, caducis. Corollæ petala 4, subrotunda, hypogyna, decidua. Stamina 25-30, hypogyna : filamenta filiformia gracilia ; antheræ terminales, oblongæ, biloculares, loculis latere longitudinaliter dehiscentibus, connectivo interposito. Ovarium cylindraceum, 5-6-sulcatum, uniloculare. Ovula numerosa, in placentas filiformes 5-6 intervalvulares demum liberas, anatropa. Stigma amplum, sessile, hemisphæricum, carnosum, ovario latius, persistens, 5-6-radiatum, radiis lamelliformibus. Capsula erecta, stricta, siliquiformis, teres, unilocularis ad apicem, infra stigma persistens, fere ad basin 5-6-valvis, valvis linearibus : placentis filiformibus liberis ad apicem stigmati unitis. Semina numerosa, ovalia, compressa, scrobiculata, strophiolata, subcristata.—Herba annua vel biennis ex Himalaya orientali, pilis longis fulvis patentibus villosa. Caulis teres, subsimplex. Folia inferiora, radicalia præcipue, longe petiolata, cordata, subpalmatim seu pedatim 5-loba, lobis lobulatis, foliis superioribus sessilibus, supremis pinnatifido-lobatis. Pedunculi terminales axillaresque. Flores cernui. Calyx hirsutus. Petala flava, magnitudine Papaveris Rhœadis. Antheræ aurantiacæ. Stigma viride.

"Found in Sikkim-Himalaya by Dr. Hooker, and reared in the Royal Gardens from seeds sent by him in the winter of 1850-1. It flowers in June, and may be treated as a hardy annual : the seeds ripening in July. The long, shaggy, fulvous hairs and bright yellow flowers give it a handsome appearance. In its foliage it differs remarkably from any of the *Papaveraceæ* with which I am acquainted, and no less in the fruit. It has the stigma of *Papaver*, while the mode of dehiscence corresponds rather with that of *Roemeria*. We cannot question its forming a new genus, which is named by Dr. Hooker in compliment to J. F. Cathcart, Esq., B.C.S., late Judge of Tirrhoot, who during a residence at Darjeeling devoted his whole time to the illustration of the botany of that neighbourhood, and superintended the execution, by native artists, at his own expense, of a collection of upwards of 700 folio-coloured plates of Himalayan plants. These drawings, which are of great botanical value, and embrace a multitude of new plants and others of the greatest beauty and rarity, are, by the liberality of their possessor, placed at Dr. Hooker's disposal for the illustration of the Botany of Sikkim. This new Papaveraceous plant was raised from seeds, received last year from the elevated regions of Sikkim-Himalaya. It appears to be a perennial rooted plant, but we must await the result of next winter, in order to know whether it is sufficiently hardy to bear the open air of this climate. Hitherto we have kept it in an airy frame, where it has flowered and produced perfect seeds. In summer it may be planted out in the open air in a cool shady place ; but at the same time care must be taken that it does not remain long saturated with moisture, for, on account of the soft and villous nature of the leaves, a continued excess of moisture may cause them to damp off."—*Bot. Mag.,* t. 4596.

384. Lilium sinicum. A handsome Chinese greenhouse bulbous plant, with scarlet flowers. Blossoms in July. Reintroduced by Messrs. Standish and Noble. (Fig. 193.)

L. sinicum; caule humili apice bi-trifloro subtomentoso, foliis sparsis oblongo-linearibus vix pubescentibus supremis sub floribus verticillatis, pedunculis nunc supra medium monophyllis, perianthii laciniis revolutis sessilibus intus lævibus circa rimam pubescentibus, staminibus perianthio brevioribus pistillo longioribus, ovario obovato obtusissimo styli longitudine.

This plant was originally imported from China by the Horticultural Society, in whose garden it flowered in September, 1824. Recently Messrs. Standish and Noble have obtained it from Mr. Fortune. It is a very dwarf kind, hardly exceeding

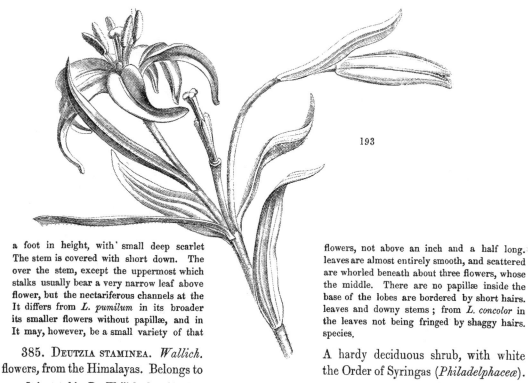

193

a foot in height, with small deep scarlet flowers, not above an inch and a half long. The stem is covered with short down. The leaves are almost entirely smooth, and scattered over the stem, except the uppermost which are whorled beneath about three flowers, whose stalks usually bear a very narrow leaf above the middle. There are no papillæ inside the flower, but the nectariferous channels at the base of the lobes are bordered by short hairs. It differs from *L. pumilum* in its broader leaves and downy stems; from *L. concolor* in its smaller flowers without papillæ, and in the leaves not being fringed by shaggy hairs. It may, however, be a small variety of that species.

385. Deutzia staminea. *Wallich.* A hardy deciduous shrub, with white flowers, from the Himalayas. Belongs to the Order of Syringas (*Philadelphaceæ*).

It is stated by Dr. Wallich that this plant grows on the highest mountains of the great valley of Nepal, and in the province of Kamaon. Dr. Royle speaks of it as being common in Mussooree, and apparently well suited to English shrubberies. It is a small bush with deciduous ovate-lanceolate stalked leaves, firmly serrated, dull-green and smooth on the upper side, whitish beneath. The flowers are pure white, somewhat larger than those of Hawthorn, in terminal corymbose panicles. The calyx is small, white, with five small triangular teeth. The petals are oblong, and rather crumpled. The stamens have large winged edges produced upwards into a strong tooth. The whole plant has a feeble somewhat balsamic smell. It is a small hardy shrub, growing well in the common garden soil, and easily increased by cuttings of the half-ripened slender young wood; is very pretty and flowers freely in May.—*Journ. of Hort. Soc.*, vol. i.

386. Grevillea rosea, *our tab.* 56. (*aliàs* Gr. lavandulacea *Henfrey.*)

Mr. Henfrey has referred this plant to the *Gr. lavandulacea* of Schlechtendahl, described in the *Linnæa*, vol. xx., p. 586, from specimens collected in South Australia by Behr; but if we are to trust the words "folia ferè teretia" and "fructus maturus extus lævis et pubescens" he can hardly be right. We must however allow that the two plants are very nearly alike, and that the supposed differences may be merely accidental. We had overlooked the paper in the *Linnæa*.

387. Fitz-Roya patagonica. *J. D. Hooker.* A noble evergreen hardy Coniferous tree from Patagonia. Introduced by Messrs. Veitch and Co.

By this name Dr. Hooker proposes to distinguish one of the most magnificent trees in Patagonia. When young, it is a graceful drooping evergreen shrub, with the habit of *Libocedrus tetragona*, to which it in fact approaches so nearly when old as not to be easily distinguishable unless in fruit. When young, the leaves are very spreading, linear, acute, decussate, narrowed at the base, flat, with two glaucous lines on the underside. When old they become triangular, sessile, closely imbricated scales, with very little appearance of glaucousness. The female flowers are little terminal stellate cones, remarkable for having the axis terminating in three soft clavate glands, or abortive scales. I have not examined them very carefully, but Mr. B. Clarke, with whose notes and sketches of this plant Sir W. Hooker has also favoured me, describes the fruit as consisting " of nine scales, three in a whorl. The lower three, which alternate with the uppermost leaves, are barren; the intermediate three only are fertile; the three uppermost alternate with the fertile and are flattened, but stand with their edges outwards. Each fertile scale has three erect seeds, surrounded by a broad wing, and ending in a narrow neck; the central seed is attached to the scale, the two lateral to the axil; sometimes two seeds are on the scale, and three on the axil." The male flowers are unknown; but as far as the females indicate distinctions, *Fitz-Roya* can be said to differ little from *Thujopsis*, except in the three terminal glands of the cone, and in three only of the scales being fertile. *Saxe-Gothæa conspicua, Fitz-Roya patagonica, Libocedrus tetragona*, and *Podocarpus nubicola* are, no doubt, the four most interesting Conifers for this country, after *Araucaria imbricata*, which South America produces.—*Journ. of Hort. Soc.*, vol. vi.

388. BERBERIS EMPETRIFOLIA; var. *cuneata*. A dwarf narrow-leaved evergreen bush, of little beauty, with solitary deep yellow flowers. Native of Patagonia, and South Chili. (Fig. 194.)

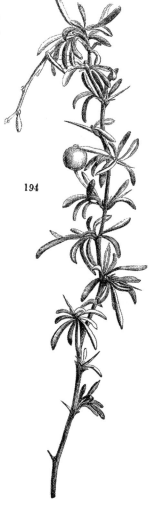

194

From the country lying between the Straits of Magellan and the Cordillera, near Valparaiso. A little trailing bush, with stiff 3-parted spines, and linear pungent leaves, not unlike those of *Genista anglica*; bright green, clustered, and about an inch long. From their axils appear, in the month of May, a few bright yellow flowers, growing singly or in pairs, on stalks shorter than the leaves. This is an humble plant, suited for rock-work in a mild climate, but among the less valuable of the genus. According to Dr. Hooker, it is confined to the Cordillera, and characteristic of a dry climate.—*Journ. of Hort. Soc.*, vol. v., p. 1.

The specimen here represented belongs to the broad-leaved variety called cuneata in the gardens.

389. CHRYSOBACTRON HOOKERI. *Colenso.* A greenhouse or frame evergreen herbaceous plant, from New Zealand. Flowers in yellow erect racemes. Belongs to the Natural Order of Lilyworts. Introduced at the Royal Botanic Garden, Kew.

The first species of the present genus (*C. Rossii*) was detected by Dr. Hooker in Lord Auckland's Islands, and it is figured and described in the *Flora Antarctica*. It was named Chrysobactron, "in allusion to the magnificent racemes of golden flowers" which that species bears. We have here a second species, far less showy, from New Zealand, whence the roots were sent by Mr. Bidwill. Mr. Colenso detected it soon after. The former gentleman found it in the rich alluvial plain of the upper part of Wairu, Middle Island; the latter in the sides of watercourses, in the country between the Ruahine range and Taupo, plentiful. "It grows in great clumps in boggy places, and is said to cover the plain with a sheet of yellow when in bloom. Some of the masses are three feet in diameter." Leaves eighteen inches long, linear-ligulate, canaliculate, glaucous-green, striated, acuminated, rather indurated at the point, the base yellowish, the three or four outer ones, nearest the root, are reduced to brown scales. Scape quite leafless, a foot and a half to two feet and even thirty inches high, erect, terete, bearing at the top numerous golden-yellow flowers in a rather lax raceme. Pedicels erect, bracteated, bracteas ovate, with a subulate point rather shorter than the pedicels. Perianth of six oblong spreading sepals. Stamens six: filaments subulate, arising from the base of the sepals. Ovary obovate, with three furrows. Style subulate, rather longer than the ovary. Capsule oblong-obovate, mucronate, elevated on a short stipes, three-celled, six-seeded. We have hitherto kept it in a cool frame during winter, for though it comes from an elevated region in a high southern latitude, we fear it may not be sufficiently hardy to bear the severity of some of our winters.—*Bot. Mag.*, t. 4602.

390. SWAMMERDAMIA GLOMERATA. *Raoul.* An insignificant evergreen half-hardy New Zealand bush, with small clusters of white flowers. Belongs to Composites. Flowers in the spring. (Fig. 195.)

This is a slender straggling naked-branched bush, with a few roundish leaves that are white with down on the underside, and a dull green, or wine-purple on the upper. According to Raoul it is found on the shore at Akaroa, in New Zealand, where it forms a bush two or three yards high. In this country it gives no sign of acquiring such a stature, but appears to be only suited for trailing over rock-work in places where the climate is mild enough for it. As an in-door plant it is not worth keeping.

391. LIGUSTRUM JAPONICUM. *Thunberg.* A hardy evergreen shrub with white flowers, belonging to the Order of Oliveworts (*Oleaceæ*). Native of Japan. Blossoms in July. (Fig. 196.)

A handsome bush, free from hairs in every part. Its leaves are oval, acute, flat, leathery, scarcely shining. The flowers are white, in loose straggling panicles. The calyxes are almost truncate, much shorter than the cylindrical tube of the corolla, beyond which the stamens project. It is very distinct from *L. lucidum*, forming a much dwarfer bush, with flatter smaller leaves, and thinner panicles of flowers. It is a good addition to hardy evergreen shrubs, for which we have to thank Dr. V. Siebold.

195

196

392. ARBUTUS VARIANS. *Bentham.* (*aliàs* A. xalapensis *Lindley;* *aliàs* A. mollis *Hooker*.) An evergreen greenhouse shrub, with panicles of white and pink flowers, and dull green leaves hoary beneath. Native of Mexico. (Fig. 197.)

This plant has been recently well figured in the *Botanical Magazine*, t. 4595, as the *Arbutus mollis* of Humboldt. It had previously found a place in the *Journal of the Horticultural Society*, v. 193, under the name of *A. xalapensis*

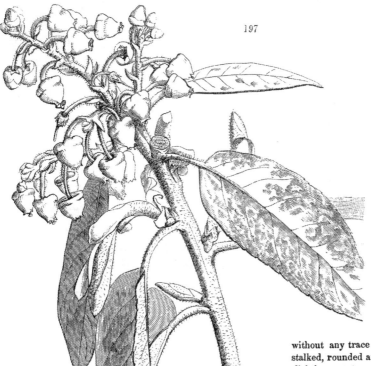

197

of Humboldt. Both Sir W. Hooker and ourselves overlooked the undoubted fact that it is the *A. varians* of Mr. Bentham in the *Plantæ Hartwegianæ*, No. 542. *A. xalapensis* seems to have a differently formed corolla, and in *A. mollis* the leaves are downy on the upper side. Sir W. Hooker thus describes the flowers :—

" Corolla large, ampullaceous or lageniform, glabrous or downy, white or greenish rose-colour ; the lower portion forms an inflated ring, the rest of the tube is hemispherical, tapering into a short contracted mouth ; limb of five small rounded lobes."

In the *Journal of the Horticultural Society* it is mentioned in the following terms :—

" Raised from Mexican seed, received from Hartweg in February, 1846, from the mountain of Anganguco. A low, dull brownish green evergreen bush. Branches, petioles, and underside of leaves covered with a short soft down, without any trace of setæ. Leaves oblong, flat, long-stalked, rounded at the base, perfectly entire, or very slightly serrate, with a hard, firm, reddish edge ; somewhat downy on the upper side. Flowers dirty reddish-white, in close downy terminal short pyramidal panicles. Peduncles glandular and woolly. Calyx nearly smooth. Corolla ovate, at the base almost flat and unequally gibbous, with a contraction below the middle, and a very small limb. Ovary with a granular surface. This little bush is by no means ornamental. It grows slowly, requires protection in winter, has dull spotted leaves, and remains in flower only for a week or two in April. Although a true Arbutus, it seems to have none of the beauty of its race, and must be consigned to the collectors of mere botanical curiosities."

We are still of opinion that the species has no horticultural value ; at least when cultivated in a cold pit it has invariably a dingy rusty aspect, the reverse of beautiful.

393. NYMPHÆA ELEGANS. *Hooker.* A charming greenhouse aquatic, with very pale violet sweet-scented flowers. Native of New Mexico. Introduced at Kew.

This has been discovered in New Mexico by Dr. Wright. Its nearest affinity, perhaps, is *N. ampla*, Bot. Mag. t. 4469. Our plants flowered in the early summer in the tank of the tropical aquarium. The blossoms are not only elegant in form and colour, but fragrant also. It will be difficult to say to which of the divisions of De Candolle this will belong. It is very different from any of the section " Cyaneæ," though its purplish-blue tint would indicate an affinity with that group. One of the most remarkable circumstances in the flower of this plant consists in the arrangement of the stamens in (apparently) as many phalanges as there are lobes to the stigma. I had not the opportunity of observing if, at a late period of inflorescence, they separated. Leaves floating, about six inches long, and four and a half or five broad, thus nearly orbicular, plane, the margin sinuated and subdentate ; the upper surface dark green, the under purple, especially towards the margin ; both sides spotted and streaked with black, the under side most spotted ; the base of the leaf is cut nearly to the petiole into two straight or slightly diverging rather acute lobes, the sinus long and narrow. Petiole terete, smooth. Scape terete, smooth, rising erect, almost a foot above the water, and bearing a fragrant flower at the top, nearly the size of our common white water-lily (*Nymphæa alba*). Calyx of four, spreading, oblong, obtusely acuminated sepals of a pale green colour, yellowish at the base, marked with numerous short streaks of deep brown. Petals twelve to fourteen, nearly of the same shape as the sepals, uniform or nearly so, yellowish-white, tinged with purplish blue. Stamens numerous, deep yellow, inner ones short and without any appendage to the anther, outer ones much larger ; the filaments broad and subpetaloid ; the anther terminated with a callous white point. The stamens in the fully expanded flower approximate in phalanges or bundles, apparently corresponding in the number of the bundles with the rays of the stigma. Ovary turbinate, bearing the petals. Stigma deep yellow, downy, about fifteen-rayed, under each ray a blunt glabrous tooth projects. —*Bot. Mag.*, t. 4604.

394. EPIDENDRUM PATENS. *Swartz.* A hothouse Epiphyte from the West Indies and Guatemala, with pale ferruginous or yellowish flowers. Introduced by G. M. Skinner, Esq. (Fig. 198 : *a*, a reduced sketch; *b*, a magnified flower.)

This, although almost unknown in collections, is probably a common West Indian plant. It grows about a foot high, with a slender stem clothed with oblong coriaceous distichous leaves. The raceme, which is terminal, is about nine inches long and is perfectly pendulous, bearing 13 or 14 flowers, of a pale rusty yellow colour, and about 1½ inch across. The sepals are thicker in texture than the petals, and somewhat darker. The lip is thin, roundish, 4-lobed, with a slight central elevated line, and a pair of thin tubercles at its base ; the lateral lobes are rounded, somewhat hatchet-shaped, and very much larger than the two in front, which are divergent. The accompanying drawing was made in the garden of the Horticultural Society. A good coloured figure of a small specimen is to be found in the *Botanical Cabinet*, t. 1537.

395. PITTOSPORUM GLABRATUM. *Lindley.* A green-flowered shrub, of little beauty, belonging to the Pittosporads. Introduced from Hong Kong. Flowers during the early spring months.

This is an evergreen greenhouse shrub, with deep-green rather blistered convex leaves, which shine, as if varnished, when young, and are somewhat glaucous underneath. The flowers appear in terminal sessile umbels, are smaller and more slender than in *P. Tobira*, of a pale-greenish white colour, and very sweet-scented. The form of its leaves and the slenderness of the corolla clearly separate it from that species.   There is also a singular tendency on the part of the leaves to assume a whorled arrangement, as in *P. cornifolium* and its allies, especially a Macao species named *P. pauciflorum*, by Messrs. Hooker and Arnott, but the calyx of that species is nearly as long as the corolla, and the petals are represented as spreading away from each other instead of being so rolled up as to resemble a monopetalous corolla.   As yet this species has been treated as a greenhouse plant, but from its appearance there is reason to believe that it may stand out of doors against a wall.   It grows freely in rough sandy peat under pot culture, but will probably succeed in common garden soil.   It strikes freely from cuttings in silver-sand under a bell-glass without much heat. Although it makes no show in a greenhouse, yet should it prove hardy, its neat foliage and sweet-scented flowers will render it a desirable plant for a conservative wall.—*Journ. of Hort. Soc.*, vol. i.

396. ANGRÆCUM ARCUATUM.   *Lindley.*   A white-flowered Epiphyte from the Cape of Good Hope.   Blossoms in July.   Introduced by Messrs. Veitch and Son.   (Fig. 199.)

The Cape of Good Hope is not the place from which we should expect to receive Epiphytes, the numerous Orchids of that country being nearly all strictly terrestrial.   Nevertheless a small number of such species are now known to botanists chiefly through the discoveries of M. Drège, an indefatigable German collector.   These plants all come from a jungly swampy district lying far to the east of Cape Town, and extending northwards at the back of Algoa Bay. There, in the district of Albany, this plant grows on trees; at a place called Kopje, on limestone hills, it also appears, growing on the roots of shrubs.   It has a stiff hard stem, from two to six inches high, clothed with tough, leathery, distichous leaves, bluntly and unequally two-lobed at the point.   The flowers, which are pure white, appear in lateral horizontal racemes, each proceeding from a broad membranous bract, which is about as long as the internodes.   The sepals, petals,

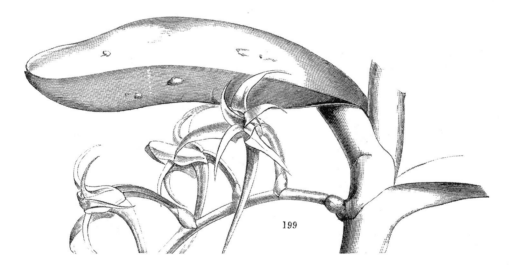

199

and lip are almost exactly alike in form, linear, taper-pointed and reflexed ; the spur is a long, tapering, blunt horn, which is much longer than the lip.   In this, as in other plants referred to the genus Angræcum, the pollen masses have each its own long narrow caudicle.

L.Constans del.& Zinc.

Printed by C.F.Cheffins,London.

[PLATE 61.]

THE CHINESE PLATYCODE.

(PLATYCODON CHINENSE.)

———◆———

A half-hardy Herbaceous Plant, from the southern Coast of CHINA, *belonging to the Order of* BELLWORTS.

═══════════════

Specific Character.

THE CHINESE PLATYCODE.—Glaucous, erect. Leaves ovate, finely serrated as far as the point. Flowers racemose. Stigmas 5. Capsule hemispherical.

PLATYCODON *CHINENSE ;* glaucum, strictum, foliis ovatis argutè serratis usque ad apicem, floribus racemosis, stigmatibus 5, capsulâ hemisphericâ.

─────────────

Platycodon grandiflorum : *Lindley in Journal of Horticultural Society,* vol. i., p. 305, not of Alphonse De Candolle.

═══════════════

THIS is the finest herbaceous plant obtained for the Horticultural Society in China by Mr. Fortune; but it requires skilful management to gain the beauty of the specimen represented in the accompanying plate, which was prepared in the Chiswick Garden. It is there cultivated in a pot, filled with peat loam and sand, the first and last in excess, exposed freely during the summer under the slight shade of a low wall, and in winter kept dry in a cold frame. Thus managed it produces fine straight stiff branches from 2 to 3 feet high, bearing several large deep blue flowers in succession at the end, and ripening seed in some abundance.

The roots are perennial, fleshy, and connected with a stout neck, where the buds are seated, from which the stems are annually produced. The latter are unbranched, glaucous, with a purplish tint, and covered with leaves from the base to the setting on of the flowers; every year they drop out of the neck (disarticulate) by a clean convex scar, which consequently leaves a concavity or socket in the neck, into which water must never be allowed to penetrate. The leaves are firm, ovate, nearly sessile, deep green above, glaucous beneath, and edged with purple; their sides are finely and equally serrated from near their base to near the point. At the ends of the shoots come the flowers, in a retrograde manner, the uppermost flower appearing first, and others afterwards in succession downwards, so that the lowest flower opens last, the inflorescence therefore being what Botanists called centrifugal. Each flower is seated on a round glaucous stalk, terminated by a smooth hemispherical ovary, bearing 5 upright narrowly triangular teeth. The corolla varies in size from 1½ inch across in ill-grown specimens, to nearly 3 inches in the most vigorous flowers; it is of a deep rich violet-blue, shaped like a balloon before expansion, and like a basin cut half way down into 5 regular sharp triangular lobes when expanded. The capsule opens at the point into 5 loculi-cidal valves, which are opposite the lobes of the calyx, the cells being consequently alternate with the lobes. The seeds are largish, black, oblong, smooth, and winged on one side.

The first knowledge we had of this plant was from finding it among some dried specimens collected by the Rev. G. H. Vachell, about the neighbourhood of Macao, and the islands adjacent, in December, 1829. Mr. Fortune brought it from Chamoo. At first we took it for a mere variety of the large-flowered Platycode, originally figured by Gmelin, from Siberia, under the name of "Campanula foliis lanceolatis glabris, inæqualiter dentatis, utroque extremo integris, ramis uni-floris terminantibus;" and under that name it has become dispersed through our Gardens. But a further acquaintance with the Chinese plant, and a comparison of it with a wild Dahurian specimen, has satisfied us that it is really quite distinct. The Russian plant is described as having a weak stem, unable to sustain itself erect (" caule surgit simplici pro ratione tenui, hinc inde flexuoso "—Gmelin), which is exactly what was found when it was formerly cultivated in our Gardens; this, on the contrary, has stiff stems, with almost a woody texture. Then the large-flowered Platycode has but one flower at the end of the stem ("in summitate flos insistit speciosissimus") or at the most two; on the contrary our species always has a long raceme, and will even sometimes branch, as is apparent from Mr. Vachell's evidence. Moreover, in the first, the capsule has the form of an inverted cone, in the last it resembles a hemisphere or half egg. We are therefore obliged to distinguish it by a new name.

There is a semi-double white variety, figured in the Journal of the Horticultural Society. Both produce seed, by which they may be propagated. Some years must however elapse before plants will bear such flowers as were produced in the specimen now represented.

L.Constans. del. & Zinc.

Printed by C.F Cheffins, London.

[PLATE 62.]

THE HYBRID CRENATE CACTUS.

(PHYLLOCACTUS SPECIOSISSIMO-CRENATUS.)

———◆———

A Garden Hybrid Greenhouse Shrub.

=================

THE following is the history of this beautiful production.   It happened that the *Phyllocactus crenatus* was in flower in the Garden of the Horticultural Society at the same time as a very fine variety of *Cereus speciosissimus* belonging to Lady Antrobus.   It occurred to Mr. Gordon to touch the former with the pollen of the latter.   In due time a fruit was formed, and *Phyllocactus crenatus* became the mother of a batch of seed which has produced the race of hybrids of which the annexed is a figure.

The seedling selected for representation is probably the finest of the crop, but all the seedlings are much alike, chiefly varying in the deeper or paler colour of their flowers.   The effect of the cross has been to form a mule with the stems and in some respects the flowers of *crenatus*, and with the colour, even as far as the well-known violet tinge, of *speciosissimus ;* so that the father gave colour and changed somewhat the form of the flowers, while the mother gave general habit.

It is evident that the Cacti mule freely.   Many are in our gardens of uncertain origin.   Sir Philip Egerton is celebrated for the success with which he has mixed them at Oulton Park, and the present case shows that great results may be thus obtained ; for this is an example of undoubted beauty.   We would, therefore, suggest the advantage of carrying these experiments much further. Why not cross the *Mammillarias* and *Echinocacti* with *Cereus* and *Phyllocactus ?*   Very singular productions might thus result.   But above all why not cross the hardy Opuntias with the brilliant species of our hothouses.   Some Opuntias will stand our winters without any protection near London, and there is no physical reason why they should not become the parents of a race of hardy and very ornamental Cacti, although they have no beauty themselves.

L Constans, del.& Zinc.

Printed by C F Cheffins, London.

[Plate 63.]

THE THREE-TONGUED ONCID.

(ONCIDIUM TRILINGUE.)

———◆———

A Hothouse Epiphyte, from Peru, *belonging to the* Orchidaceous *Order.*

Specific Character.

THE THREE-TONGUED ONCID. Raceme somewhat twining, panicled at the base. Flowers thin. Bracts oblong, spathaceous, four times shorter than the ovary. Lateral sepals unguiculate, connate at the base, lanceolate, long, wavy; that at the back roundish ovate, crisp, the claw eared at the base and as long as the column. Petals lanceolate, revolute, very crisp. Lip dagger-shaped, crisp, revolute; the segments at the base coarsely toothed, fleshy, ascending, with a very large convex crest, three-tongued in front, having two tubercles behind, a thin plate lying between, and a fleshy tooth on each side. Column smooth, with small bristle-shaped wings.

ONCIDIUM *TRILINGUE,* (Microchila) foliis . . ., racemo subvolubili basi paniculato, floribus raris, bracteis oblongis spathaceis ovario quadruplò brevioribus, sepalis lateralibus unguiculatis basi connatis lanceolatis undulatis elongatis dorsali subrotundo-ovato crispo ungue auriculato columnæ longitudine, petalis lanceolatis revolutis valde crispis, labelli pugioniformis crispi revoluti auriculis grossè dentatis carnosis ascendentibus cristâ maximâ valdè convexâ à fronte trilingui à tergo bituberculatâ laminâ tenui interjectâ denticulo carnoso utrinque, columnæ glabræ alis parvis setaceis.

Oncidium trilingue : *suprà,* vol. i., p. 42, no. 63.

THIS curious plant has already been noticed in our work at the place above mentioned. Since then good fresh specimens, from Sir Philip Egerton, have enabled our artist to produce a coloured figure of it. It is a species of Oncid, with the habit of *O. macranthum,* but with flowers quite unlike anything in our gardens. It is, however, associated in nature with many species of similar habit, having a small fleshy lip, combined with large and unusually unguiculated sepals; they are the Cyrtochils of Humboldt and Kunth, and form a complete transition to the genus Odontoglossum, from a portion of which they in fact differ in nothing except the lip having no adhesion to the face of the column.

The Cyrtochilian Oncids, as these plants might be termed, comprehend eleven certain, and two doubtful species, all from the tropical parts of South America, where they grow on trees, and produce long rambling panicles of large brownish flowers variously mottled with yellow and purple; not unfrequently these panicles twine round the neighbouring branches, a property which seems essential to them in order that their heavy flowers may be supported. The species now figured, which gives a good idea of the habit of many of them, is perfectly distinguished by its short crisp petals, and its singular lip, the callosities upon which are not easy to represent either by words or a drawing. In form the lip represents a long trowel, curved inwards at the edge, and backwards at the point; near its base is planted a pale yellow plate, free at the edges, but extended in the middle into three tongue-like yellow processes (a section of it in front has the appearance of fig. A); near the base the section is like fig. B, in consequence of the yellow plate above described becoming more free from the lip. Towards the base of the three tongues rises on either side a small purplish tubercle; two others stand on the lip at its base; and between the two pairs of tubercles the three tongues rise up into a white triple tooth.

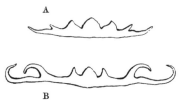

The following memoranda will serve to guide both botanists and collectors to a knowledge of this curious section of Oncidium.

ONCIDIA MICROCHILA—Cyrtochilum H.B.K. L. p. 210.

1. O. trifurcatum *Lindl. in Ann. Nat. Hist.*, vol. xv.; sepalis lateralibus unguiculatis spathulato-obovatis planis dorsali unguiculato rotundato crispo duplò breviori, petalis oblongis crispis dorsali minoribus, labello unguiculato tripartito laciniis linearibus truncatis lateralibus canaliculatis, crista trilamellata, columna tetraptera alis superioribus linearibus carnosis apice abrupte recurvis inferioribus rotundatis tenuioribus, clinandrii dorso in dente antherifero producto, rostello membranaceo bifido. —*Peru* (Hartweg).—I have only seen three flowers of this. They are three inches in diameter; the lateral sepals are whole-coloured, the dorsal and the petals are bordered with yellow (?). It stands near *O. serratum.*

2. O. serratum *Lindl. Sert. Orch.*, sub t. 48; *suprà*, vol. i., no. 42, fig. 15; pseudobulbis ovalibus diphyllis, foliis erectis rigidis acutis basi angustatis canaliculatis paniculâ pauciflorâ brevioribus, sepalis serrato-crispatis dorsali reniformi lateralibus multò longioribus obovatis patentissimis, petalis ovatis acutis serrato-crispatis conniventibus, labello multò minore hastato: laciniis acutis intermediâ lineari obtusâ medio constrictâ lateralibus acuminatis 3-plò minoribus, (cristâ depressâ crenulatâ), columnæ alis subulatis ascendentibus.—*Peru.*—This singular plant has large flowers, brown, oblong, smooth, terete pseudobulbs, each having two broad sword-shaped leaves at the point, and several others below the pseudobulbs. The flower-stem is nine feet long, partly twining, with five or six lateral branches, each carrying from four to six flowers near the extremity. These flowers are said to be cinnamon-brown in Peru, with bright yellow tips to the upper divisions. In the fresh flower they have the colour of *Oncidium luridum*, only brighter; but the yellow on the upper half of the delicately fringed and crisped petals is clear and brilliant. It flowered with M. Pescatore at Paris.

3. O. macranthum *L.* p. 205; pseudobulbis ovatis, foliis oblongis obtusis, racemo volubili, sepalis cordatis oblongis obtusis undulatis unguiculatis, petalis æqualibus conformibus paulò latioribus et breviùs unguiculatis, labelli hastati laciniis lateralibus subfalcatis intermediæ acuminatæ æqualibus cristâ trilamellatâ; lamellis apice confluentibus utrinque dentibus duabus runcinatis, columnæ alis rotundatis.—*Guayaquil.*—Flowers three or four inches across. Sepals purplish-brown, tipped with yellow. Petals bright yellow. Lip purple, with a white crest.

4. O. cordatum *Lindl. Sert. Orch.*, sub t. 25; pseudobulbis . . ., foliis oblongo-lanceolatis acutis coriaceis basi angustatis, scapo paniculato ramosissimo, bracteis oblongis cucullatis membranaceis obtusis, sepalis unguiculatis ovatis undulatis, petalis unguiculatis cordatis margine crispis denticulatis, labelli hastati unguiculati lobis angustis acuminatis appendicibus disci petaloideis, columnâ subapterâ.—*Peru ; rocks on the road to Pangoa.*—A fine large species, with brown flowers whose segments appear to be bordered with yellow.

5. O. falcipetalum *Lindl. Orch. Linden.*, no. 76; foliis lanceolatis acutis pergameneis 7-9-nerviis, floribus densè paniculatis, bracteis cymbiformibus obtusis, sepalis brevè unguiculatis supremo subrotundo-cordato lateralibus ovato-oblongis obtusis, petalis sessilibus brevioribus oblongis crispis complicatis falcatim recurvis, labello carnoso hastato acutissimo basi utrinque corrugato et lamellis dentatis aucto, columnâ lævi, alis parvis semiovatis unidentatis, clinandrio postice mucronato pubescente.—*Both epiphyte and terrestrial, from the forests of Merida, at the height of* 5000 *to* 6000 *feet.*—Pseudobulbs oval, compressed, two or three inches long. Stem twenty feet long, scrambling. Flowers very large, brown. A noble species, with flowers more than three inches in diameter. Leaves eighteen inches long, one and a half wide.

6. O. trilingue *Lindley in Paxton's Fl. Garden*, vol. i., no. 63; vol. ii., t. 63; foliis . . . . racemo subvolubili basi paniculato, floribus raris, bracteis oblongis spathaceis ovario quadruplò brevioribus, sepalis lateralibus unguiculatis basi connatis lanceolatis undulatis elongatis dorsali subrotundo-ovato crispo ungue auriculato columnæ longitudine, petalis lanceolatis revolutis valdè crispis, labelli pugioniformis crispi revoluti auriculis grossè dentatis carnosis ascendentibus cristâ maximâ valdè convexâ à fronte trilingui à tergo bituberculato laminâ tenui interjectâ denticulo carnoso utrinque, columnæ glabræ alis parvis setaceis.—*Peru.*—Flowers cinnamon-brown, in a long half-climbing panicle. Lip dagger-shaped, with a yellow crest, consisting of three flat yellow tongues terminating a thin winged plate.

7. O. superbiens *Reichenbach fil. in Linnæa*, vol. xxii., p. 843; "foliis oblongo-lanceolatis, acutis, latis, paniculâ maximâ, bracteis cymbiformibus acutis, sepalo supremo reniformi, unguiculato, basi utrinque auriculato; lateralibus obtusè triangularibus longius unguiculatis, basi pariter auriculatis; petalis subæqualibus, sed brevius latiusque unguiculatis, undulatis; labello triangulari, brevissimè unguiculato, basi utrinque obtusato, apice acuminato, callo cristæformi antice 3—5-dentato in disco, dentibus 2 lateralibus in basi; columnæ alis erectis, retrorsum falcatis, androclinio postice dentato."— *N. Grenada, near Pamplona.*—Discovered by Funk and Schlim in January, 1847. Flowers brown; the lateral sepals yellow, with brown spots. Said to be near *O. halteratum.*

8. O. halteratum *Lindl. Orch. Linden.*, no. 75; foliis ensiformibus tenuibus acutis, racemis laxis longissimis, bracteis cymbiformibus obtusis, sepalis longè unguiculatis supremo cordato-subrotundo lateralibus ovatis obtusis, petalis sessilibus ovatis undulatis obtusis brevioribus, labello carnoso lanceolato acuto subtus carinato suprà cristâ elevatâ etiam carinatâ pubescente aucto basi utrinque dentato, columnæ dorso sub apice glanduloso, alis linearibus retrorsum falcatis.—*Epiphyte from*

the forests of Quindiu, in the province of Maraquita, at the height of 7800 *feet; February.*—Flowers deep yellow. A very fine species. Flowers more than two inches across. Leaves thin, scarcely an inch broad.

9. O. undulatum *Lindl. Sert. Orch.*, sub t. 48; (*Cyrtochilum undulatum* H. B. K., L. p. 210); "foliolis calycinis ovatis undulatis, patentibus."—*N. Grenada.*—Flowers brown, spotted with white and yellow. Lip pink outside, yellow within, variegated with red and white spots. Scape as high as a man, according to Kunth. Quite indeterminable without access to the original specimen.

10. O. flexuosum *Lindl. Sert. Orch.*, sub t. 48; (*Cyrtochilum flexuosum* H. B. K., L. p. 210); "foliolis calycinis undulatis reflexis, exterioribus spathulatis, interioribus obovatis."—*N. Grenada.*— Lip ovate, acute, convex, crested with tubercles at the base. Scape several feet high, much branched, with triangular ramifications, according to Kunth. A mere puzzle without access to the original specimens.

11. O. corynephorum *Lindl. Sert. Orch.*, sub t. 25; (*Cyrtochilum volubile* Pœppig nov. gen. &c. 1. 35. t. 61.); pseudobulbis angustissimis compressis, foliis angusto-lanceolatis acutissimis, scapo ramoso paniculato, bracteis membranaceis subrotundis obtusissimis, sepalis subrotundo-oblongis longè unguiculatis, petalis angustioribus lanceolatis acutis reflexis, labello sessili obovato rotundato : callis baseos depressis apice trinis latere rugosis tuberculatis, columnâ clavatâ alis inflexis.—*Peru.*—The twining scapes are from 15 to 20 feet long. Flowers two inches in diameter. Sepals violet. Petals white, tinged with rose. Lip deep crimson above the middle. Notwithstanding the difference between this character and Pœppig's barbarous figure, I have no doubt it is the same plant as his.

12. O. loxense, *sp. nov.;* paniculâ ramosâ divaricatâ ramulis 2-3 floris, bracteis brevibus ovatis obtusis, sepalis oblongis planis reflexis petalisque paulò latioribus apice rotundatis, labello subrotundo basi sub-hastato apice excavato, callis baseos 3 parallelis ramentis pluribus à fronte, columnâ apterâ basi bibrachiatâ.—*Cordillera near Loxa, flowering in July.*—Of this a single plant was found by Hartweg, with a flower-stem 9 feet long. It is very near *O. corynephorum* but the flowers are more than twice as large, the lip has quite a different form, and the column has two short spreading arms near the base, of which no trace is to be found either in Pœppig's figure or in Mathews' drawing in our possession.

13. O. microchilum *Bateman in Bot. Reg.* 1840, misc. 193—1843, t. 23; pseudobulbis lenticularibus brevibus monophyllis, folio oblongo carinato carnosissimo acuto quam scapus erectus versus apicem paniculatus quadruplò breviore, sepalis liberis lateralibus longiùs unguiculatis, petalis oblongis subundulatis retusis, labello duplò latiore quam longo lobo intermedio nano triangulari. lateralibus rotundatis planis, cristâ reniformi crenatâ, columnæ nanæ alis subulatis apice glandulosis. —*Guatemala.*—Flowers in a large branching glaucous panicle. Sepals dull brown. Petals dull purple, with a yellowish border. Lip spotted, crimson and yellow in the centre, pure white on the side segments. Wings of column yellow, pointed with purple.

GLEANINGS AND ORIGINAL MEMORANDA.

397. MIMOSA URUGUENSIS. *Hooker and Arnott.* A half-hardy handsome spiny shrub, with brick-red flowers. Native of the Banda Oriental. Blossoms in June, July, and August. (Fig. 200.)

This shrub was originally raised in the garden of the Horticultural Society, from seed obtained from Buenos Ayres by the Hon. W. F. Strangways ; and first flowered at Chiswick in June, 1841. Since that time it has been cultivated in the same establishment, though scarcely known elsewhere, and proves to be a useful shrub for the summer decoration of borders, where it lives and flowers freely till the approach of winter. It is about as hardy as the general mass of New Holland Acacias. The flowers have a light and elegant appearance, with their reddish or brickdust tint, among the finely divided shining delicate foliage. The branches are always short, and furnished with little straight spines, each branch bearing from two to four balls of flowers. The chief horticultural defect of this plant is that, like other hard-wooded small-leaved shrubs, it is rather too thin of leaves, and becomes naked at the base when old. When we find a plant like this, of a tropical genus, so nearly hardy, we can but entertain a confident opinion that the countries watered by the River Uruguay deserve to be visited by some horticultural collector.

398. CATTLEYA LEOPOLDI. *Hort.* A beautiful stove Epiphyte with brownish yellow spotted flowers, and a rich crimson lip. Native of Brazil.

This is a mere variety of *Cattleya granulosa,* with a most brilliant tint of rich purplish crimson in the lip. It is one of the handsomest orchids in cultivation, and seems to have reached us through the Belgians, the first we heard of it being that it had been exhibited at Brussels by Mr. Forkel, gardener to King Leopold at Laeken.

399. CLEMATIS HEXASEPALA. *De Candolle.* (*aliàs* C. hexapetala *Forster.*) A half-hardy green-flowered twining fragrant plant, belonging to the Order of Crowfoots. Native of New Zealand.

This is a little twining plant, with shining nearly smooth ternate or biternate leaves, whose petioles twine round any small body with which they may come in contact. The leaflets are cordate-ovate, coarsely serrated, and often three-lobed. The flowers are small, pale green, very sweet-scented, and appear in threes or fours from the axils of the leaves. Their stalks are long and hairy, and each has a pair of small bracts below the middle. The sepals are very uniformly six in number, of a narrowly oblong form, and spreading so as to form a small green star. Contrary to the usual structure of the genus, the stamens are constantly six only in number, and about half as long as the sepals. The late Mr. Allan Cunningham gathered it in the northern island of New Zealand, but it was first found by Sir Joseph Banks in 1769, and a drawing of it is said to be preserved in the Banksian Library. It is a hardy greenhouse plant, requiring a light loamy soil to grow in, and is easily increased by cuttings of the half ripened wood. It only requires the protection of a cold pit or frame during winter, and flowers abundantly in April. Although its blossoms are green and inconspicuous, it is far from an unimportant species, on account of its blooming freely, and being very sweet-scented.—*Journ. of Hort. Soc.*, vol. i.

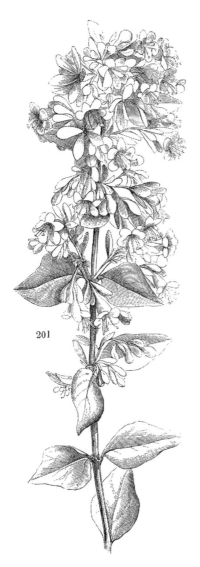

201

400. ABELIA RUPESTRIS. *Lindley.* A fine dwarf shrub, found amongst rocks on the Chamoo Hills of China. Flowers white. Belongs to Caprifoils. (Fig. 201.)

A small spreading bush, with deciduous, bright green foliage. The branches are very slender, covered with fine down, and deep reddish brown, when fully exposed to the sun. The leaves are opposite, ovate, distantly serrated, on very short stalks, quite smooth except at the midrib on the underside, where they are closely covered with short hairs. The flowers are pure white, something like those from the honeysuckle, and come in pairs from the axils of leaves belonging to the short lateral branches. At the base of the ovary stand three very small bracts. The ovary itself is slender and downy; surmounted by a calyx of five obovate ciliated sepals, which are slightly stained rose-colour, and rather membranous. The corolla when expanded is half an inch long, funnel-shaped, downy with a spreading border of five convex ovate blunt equal lobes, beyond whose tube extend four smooth filaments. The plant is distinguishable from *Abelia chinensis* of Brown, by its want of involucre, smooth leaves, and not trichotomous flowers; and from the *Abelia serrata* of Zuccarini and Siebold, by its 5-leaved calyx. It has hitherto been treated as a greenhouse plant, but will probably prove hardy enough to stand out of doors in mild winters. The soil which appears most suitable is rough sandy loam, mixed with a little peat. Being of free growth, an ample supply of water is necessary during the summer season. In winter nothing different from the general treatment of greenhouse plants is required. It is propagated from cuttings of young wood, in the usual way. From its being sweet-scented, and the length of time it remains in flower, this will be of considerable importance as a greenhouse plant; and should it prove hardy, it will doubtless be a good addition to the shrubbery in consequence of its flowering in autumn. [The foregoing remarks were made in the *Journal of the Horticultural Society* soon after the introduction of this plant. We have now to add, that although a most useful greenhouse plant, it does not prove hardy enough for the open air in the neighbourhood of London.]

401. OPHIOPOGON PROLIFER. *Lindley.* A white-flowered hothouse perennial. Native of Penang. Belongs to Lilyworts (*Liliaceæ*).

This is an evergreen herbaceous plant, with a slender stem slowly rising by means of roots which its leafy stems throw out, in the manner of a screw pine. The stems are not thicker than a swan's quill, and bear at intervals clusters of bright green sword-shaped leaves, which curve downwards and are longer than the flowering stems. The latter are bright purple, and bear in an interrupted manner a few clusters of nearly sessile small white obovate flowers, whose texture is between fleshy and spongy. In this species the stamens are united in a very short fleshy ring. The ovary is very thin-skinned, and adheres, but does not grow, to the perianth. In each of its three cells stands a pair of fleshy ascending anatropal ovules. The style is pyramidal and terminated by three small point-like stigmas. It is a stove plant, which appears to succeed well in rough sandy peat. During summer an ample supply of water is necessary; also a very moist atmosphere, at a temperature of not less than 80° by day. In winter it requires to be treated almost like an orchidaceous plant : if a humid atmosphere is kept up, little or no water will be required for a few weeks. It appears to be an abundant flowerer, and is of some interest to those who delight in curious stove-plants.—*Journ. of Hort. Soc.*, vol. i.

402. CAPSICUM CEREOLUM. *Bertoloni*. A very pretty South American half-shrubby plant, with bright yellow waxy fruit. Belongs to Nightshades. Introduced by Messrs. Veitch and Son. (Fig. 202.)

We presume that this is the plant which Professor Bertoloni thus named in his account of new plants in the Garden of Bologna ; but we have never seen that work. At all events the name is a happy one, and the short definition in Walpers is strikingly applicable. It is a very neat-looking pale-green half-shrubby plant, with oval leaves always tapering to the point, and at the base sometimes rounded, sometimes acute and oblique. They are downy all over, especially at

202

the midrib on the underside where they are woolly. The fruit is curved backwards, conical, very sharp, a little contracted in the middle, of a clear bright lemon-yellow colour. The calyx of the fruit is circular (or truncated), with five obscure very short teeth. It is quite distinct from any of the species before in cultivation, and forms a gay and useful ornament of the greenhouse during summer and autumn. *C. cereolum* is said to be from Brazil ; this is we believe the result of Mr. Lobb's collections on the west coast of South America.

403. ARAUCARIA COOKII. *R. Brown.* A very large greenhouse Coniferous tree from New Caledonia. Introduced by Mr. C. Moore.

In the year 1850 Mr. Charles Moore, the Superintendent of the Botanic Garden, Sydney, was enabled to pay a short visit to New Caledonia and the neighbouring islands of the South Pacific, in H.M.S. "Havannah;" and, notwithstanding many difficulties, succeeded, through the very great kindness of Captain Erskine, in collecting and bringing safe to Sydney a considerable number of very valuable plants, seeds, and specimens. Some of them have been brought to England by Captain Jones, of the "St. George" merchantman ; and among them the plant at the head of this article, which grows abundantly on the islands of Aniteura, New Hebrides, and New Caledonia. In a memorandum that accompanied the plant received by the Society, Mr. Moore remarks that the tree is "apparently distinct from *A. excelsa.* It differs from that species in having a more compact habit when old, and in being less rigid and more graceful when young, in the scales of the cone having a longer and more reflexed mucro, and in their gibbous, not wedge-shaped form, as in *A. excelsa.* In the island of Aniteura this plant has become scarce, the English traders having cut it down for ships' spars. I only saw one plant, and this was 'tabooed,' or rendered sacred, by the natives ; but in New Caledonia, on the south-east coast, whole forests composed of this alone were observed. In such situations the tops are not unlike basaltic columns, and were actually taken for such by the naturalists who accompanied Cook. A coral reef connects the Isle of Pines with that part." Mr. Moore adds, that it is "singular enough the first plant of this, noticed by Cook (described by that navigator, in his account of New Caledonia, 'as an elevation like a tower'), still stands, and is in a flourishing condition. Its appearance now is exactly that of a well-proportioned factory chimney of great height. The cone shows how very distinct this is from either *A. excelsa* or *Cunninghamii.* In addition to the greater length of the reflexed appendages on the scales of *A. Cookii,* to which Mr. Moore has drawn attention, it is to be observed that the scales themselves do not terminate in a hard, woody, truncated extremity, as in those two species, but are wholly surrounded by a thin wing ; the effect of which is to destroy the knobby appearance of their cone, and to give it a softness and evenness peculiar to itself."—*Journ. of Hort. Soc.,* vol. vi.

404. CALANTHE VERATRIFOLIA. *R. Brown;* var. *australis* (*aliàs* C. australis *Hort.*) A greenhouse terrestrial Orchid from New Holland, with white flowers changing to buff. Flowers in September. Reintroduced by Mr. C. Moore, Superintendent of the Botanic Garden, Sydney.

This plant was originally found by the late Allan Cunningham in August 1822 ; whilst on an excursion to the Illawarra, a coast district on the south of Port Jackson, he met with a plant in dark shaded woods, which he introduced to Kew in the following year, considering it a Bletia. It soon afterwards flowered in that collection, and was then ascertained to differ in no material respect from the plant of the Indian Archipelago. Both have been in flower together, and on examination of the two plants, no difference has been discovered, excepting that the Australian plant is not so purely white in the flower as the one from India. Mr. D. Moore of the Glasnevin Gardens, who has recently received live plants from his brother, is of opinion that "the spur is shorter, and the flowers more compact than those of *C. veratrifolia.* The leaves are also shorter and the plant is hardier, having stood in a cool greenhouse all the season and flowered nicely." We cannot however say that the flowers with which Mr. Moore has favoured us exhibit any appreciable structural difference, and we must therefore continue to leave the plant as a mere geographical variety.

405. COTONEASTER THYMIFOLIA *of Gardens.* A small prostrate evergreen hardy shrub from Gossain Than. Belongs to Appleworts (*Pomaceæ*). Introduced from France. (Fig. 203.)

It is certain that this curious little evergreen shrub is a mere variety of *Cotoneaster microphylla,* next to *C. rotundifolia,* the most beautiful of the Indian Cotoneasters. There appears to be no distinction between the two, nor any difference beyond size. *C. thymifolia* is not half the size of *C. microphylla,* lies flat on the ground like thyme itself, or if upon a stone hardly raises its head above the surface ; its leaves are not more than a quarter the size, and are much narrower in proportion, but they have the same texture, surface, point, and hairiness underneath ; they are not so generally emarginate, though they are sometimes so ; the fruit is much smaller, and so are the petals ; it seems to be identical with the Gossain Than specimens distributed by Dr. Wallich under the number 662 of his Herbarium. For rock-work, or similar places, it is quite a little acquisition. For the purpose of placing it securely on record, we add a short technical phrase which will enable it to be distinguished, whether as a species or mere variety :—

203

C. thymifolia ; sempervirens, lucida ; ramis prostratis intertextis, foliis lineari-obovatis obtusis margine recurvis subtùs pubescentibus, pomis subsessilibus solitariis recurvis, petalis inconspicuis.

204

406. DENDROBIUM GIBSONI. *Paxton.* A beautiful Epiphyte from the jungles of India. Flowers rich apricot-yellow, with a purple stain on the lip. Blossoms all the summer. Introduced by the Duke of Devonshire. (Fig. 204.)

Dendrobium (Stachyobium) *Gibsoni ;* foliis acuminatis, racemis nutantibus pendulisque multifloris elongatis, bracteis minutis ovatis obtusis, floribus subcarnosis, sepalis subrotundis basi in cornu brevi connatis, petalis latioribus integerrimis, labello cochleato cucullato obtuso villoso fimbriato.

We do not find a description of this noble plant, common as it is in gardens, under a name given to it some years since by one of us, in compliment to the industrious collector by whom it was first transferred from India to Chatsworth. Nevertheless it is one of the finest of the yellow species, rivalling even *D. clavatum* in brilliancy ; from that species it differs manifestly in its obsolete bracts and much smaller flowers. Its nearest ally is *D. fimbriatum,* with which we have reason to think it is sometimes confounded. Like *D. fimbriatum,* it bears its flowers in long nodding racemes ; their colour is yellow, and they are stained on the inside with a purple blotch in the same manner as those of the plant figured in the *Botanical Magazine* under the name of *D. fimbriatum oculatum.* But the flowers are smaller, between fleshy and leathery in texture, much blunter in the bud, in consequence of the greater roundness of all the parts, and the petals are entirely destitute of the fringe which accompanies those of *D. fimbriatum.*

We find it in our Herbarium from Griffith, gathered on Mango-trees in the province of Tenasserim, with the following note :—" Flores aurei ; labellum cochleato-cucullatum, pulcherrimè fimbriatum, cucullo rubro striato, maculâ atrosanguineâ ad ejus orificium." The specimen here represented was produced at Chatsworth last September. At the same time we saw it covered with flowers in Messrs. Veitch's magnificent Orchidhouse at Exeter.

407. DESFONTAINEA SPINOSA. *Ruiz and Pavon.* A hardy (?) evergreen shrub, with long

tubular crimson and yellow flowers.   Native of Patagonia.   Natural Order uncertain.   Introduced by Messrs. Veitch & Co.

There is so much resemblance between this plant and a common Holly, that if its leaves were not opposite, it might be mistaken for one when not in flower.   Its blossoms, however, of which one has been produced in Messrs. Veitch's Nursery, are almost 2 inches long, cylindrical, with a scarlet tube and a yellow border.   As it naturally produces a great abundance of these brilliant blossoms, it must be a most charming plant when in fine condition.   According to Dr. Hooker, the plant extends to the Andes under the equator, at the elevation of 12,000 feet, to the level of the sea, in Staten Island, in latitude 53° south.   According to Mr. Lobb, it seldom grows more than 5 feet high ; and, from the places in which it is found, he thinks it may be hardy.   It will be better, however, to consider it, in the first instance, a plant that requires protection in winter.—*Journ. of Hort. Soc.*, vol. vi.

408. PERNETTYA CILIARIS. *Don.* A hardy evergreen shrub from the mountains of South Brazil. Belongs to Heathworts.   Berries rich deep purple.   Introduced by Messrs. Veitch and Co.

In the nursery of Messrs. Veitch there grows in the open air a dark-green low bush, with hard evergreen, ovate, serrated, wrinkled leaves, covered slightly with stiff brown hairs on the under side.   The branches are clothed with similar hairs.   In appearance it is not unlike *Vaccinium Arctostaphylos*.   The flowers grow in numerous erect dense racemes, and are succeeded by piles of deep rich, reddish brown, depressed umbilicate berries, with a smooth calyx, the base of whose sepals is gibbous, fleshy, and hairless.   The stalks are, however, hispid, and about twice as long as a smooth, pale, cucullate bract, which wraps round their base.   The bush is said to have been obtained from Brazil, but it appears to agree altogether with the *Pernettya ciliaris* of Don, said to be from Mexico, of which I have seen no specimen in the many collections from that country.   Mixed with the bright rosy berries of *P. mucronata* and *angustifolia*, this produces a very gay effect in the American border.—*Journ. of Hort. Soc.*, vol. vi.

409. PRIMULA INVOLUCRATA. *Wallich.* A handsome, hardy, herbaceous plant.   Flowers white.   From the North of India.   (Fig. 205.)

When at rest this plant forms a large egg-shaped bud, which may almost be called a bulb.   Early in the spring it throws up a tuft of smooth shining leaves, the colour and texture of *Pilewort*, which are immediately succeeded by a scape from six to nine inches high, terminated by three or four white, sweet-scented flowers, which are at first slightly yellow, and when dying acquire a tinge of blush. In form, the leaves are long-stalked, ovate, obtuse, wavy, and slightly toothed.   The involucre is remarkable for having its base extended downwards into a sheath, in the same manner as in *Thrift*.   The corolla is about the size of a cowslip, with a flat border, whose segments are round and two-lobed, and a pink tube which is a little longer than the angular calyx.   It is a hardy perennial, growing about six inches high, in a soil composed of sandy-loam and leaf-mould.   It will flower in the open border about March, but earlier if kept in a cool greenhouse or frame.   It is stated by Capt. Munro that he collected it at an elevation of 11,500 feet, growing in the neighbourhood of water.   It will be a most desirable little plant for rockwork not too much exposed to a hot sun.—*Journ. of Hort. Soc.*, vol. i.

205

410. EUCRYPHIA CORDIFOLIA. *Cavanilles.* A very fine evergreen hardy (?) shrub, with broad sessile heart-shaped leaves and large axillary flowers.   Native of Chiloe and Patagonia. Belongs to Tutsans (*Hypericaceæ*).   Introduced by Messrs. Veitch & Co.

We lately saw this noble plant growing in the open air in Messrs. Veitch's Nursery.   It has a stiff hard-wooded habit, with downy branches.   The leaves, which sit close together on the stem, are hard, like those of an evergreen oak, from 2 to 3 inches long, dark green, oblong, nearly sessile, heart-shaped, with shallow toothings at the edge ; on the under side they are covered with a close short felt. The flowers, unknown in a fresh state in this country, appear to be white, are about as large as a small Camellia, and grow singly in the axil of the uppermost leaves. In his correspondence, Mr. Lobb, who sent the plant to Messrs. Veitch, speaks of it thus :—" The Eucryphia is much like *Quercus Ilex*, and I think will prove hardy. When I left San Carlos it was in full bloom. It is the most showy tree of the country. The hardiness of plants greatly depends on the nature of the wood : for instance, the *Eucryphia cordifolia* is a hard-wooded tree, and would probably stand the winter without injury ; while those of a soft-wooded nature, such as *Drymis chilensis*, *Laurus aromatica*, and others from the same locality, would be much injured if not killed."

411. SIPHOCAMPYLUS AMŒNUS. *Planchon.* A fine greenhouse shrub from the mountains of Brazil. Flowers rich orange red. Belongs to Lobeliads. (Fig. 206.)

M. Planchon states that this beautiful thing was raised from the earth of a parcel of Orchids sent from Brazil by M. Ghiesbreght. It flowered in the garden of the King of the Belgians at Laeken. The plant is described as more herbaceous than shrubby, with erect, angular, and rather downy branches. The leaves are oblong-lanceolate, bright green, with glandular serratures, having a silky lustre on the upper side, and very minute down on the under. The numerous flowers are arranged in one-sided racemes, are small for the genus, and of a rich orange red.—*Flore des Serres.*

412. LAPAGERIA ROSEA. *Ruiz and Pavon.* A greenhouse climber, with very large pendulous flowers, rich purple, a little mottled with white. Native of Chiloe. Belongs to Philesiads. Introduced by G. T. Davy, Esq.

A climbing plant from the south of Chili : it is of large growth, and scrambles over bushes in the woods of Chiloe, producing there firm, broad, dark-green leaves, and brilliant, rose-coloured, speckled, pendulous, campanulate flowers, as large as a tulip. In a conservatory where the roots have plenty of room to spread it has flowered with Messrs. Veitch, but is a plant of very difficult management. It would be a great gain to gardens if this plant would prove hardy. Such experience, however, as has been gained is unfavourable to the supposition. Nevertheless, Mr. Lobb is of a different opinion, as will be seen by the following extract from his letters :—" Respecting the hardiness of these things (*Lapageria rosea, Luzuriaga radicans,* and *Callixene polyphylla,*) if you look at their geographical position, it may be assumed that all from the elevated parts of the mainland are hardy, and I think that those from the low grounds will only require sheltered situations. The climate of Chiloe is much like that of Cornwall ; it rains almost incessantly in the winter months, but it is never so cold in winter as it is in England. Frost often occurs, but of short duration. Summer is also wet and cold ; the thermometer seldom rising beyond 65° ; but although the frost is not so severe, the south winds are very cold and cutting, and I am inclined to think that, if any thing be required, it will be sheltered situations for those that inhabit the low grounds near the sea."

206

413. PRIMULA SIKKIMENSIS. *Hooker.* A yellow-flowered Primrose from Sikkim-Himalaya, with something the appearance of an Oxlip. Flowers in May. Introduced at Kew.

" Among the drawings sent home by Dr. Hooker from Sikkim-Himalaya, is one of a yellow Primula of which that traveller relates, 'It is the pride of all the Alpine Primulas, inhabits wet boggy places at elevations of from 12-17,000 feet, at Lachen and Lachong, covering acres with a yellow carpet in May and June.'" It is, perhaps, the tallest Primula in cultivation, and very different from any hitherto described. Stemless. Leaves all from the root, erecto-patent, 8-9 inches to a foot long (including the petiole), obovato-oblong, thin and submembranaceous, but strongly reticulato-venose, not farinose, obtuse, the margin doubly and sharply toothed, the thickened midrib and nerves prominent beneath, where the hue is paler than above; they taper into a long broad red petiole about equal in length to the leaf. Scape a foot to two feet high, erect, terete, pale green, bearing an umbel of lemon-yellow (rather than golden) flowers, about the size of those of *P. vulgaris*. Involucre of 5-7 leaflets, which are sessile, slightly farinose, erect, lanceolate, a little tinged with red, about half the length of the pedicels. These latter are slightly spreading. Calyx tinged with purple, farinose, tubular-oblong, as long as the tube of the corolla, five-lobed about half-way down, lobes erect, rather obtuse. Corolla with the tube as long as the calyx, the limb subcampanulate, the mouth being wide, not at all contracted, naked, the lobes of the limb moderately spreading, roundish, emarginate. A free-growing species, partaking of the habit of the common Primrose, and therefore more permanent under artificial cultivation than the fugacious *Primula capitata* from the same country. During the winter we kept the young plants under the protection of a frame; and we shall not know, until next winter has passed, whether this species is sufficiently hardy to withstand, unprotected, the cold of our winters.—*Bot. Mag.*, t. 4597.

414. FICUS VIRGATA. *Roxburgh.* A half-hardy deciduous shrub, with broad rough leaves. Native of the North of India. (Fig. 207.)

In general appearance this shrub resembles the common eatable fig, but it seems to form a much smaller bush. The young shoots, leaves, and fruit are covered with fine short hairs. The leaves are roundish-ovate, from three to four inches long, regularly serrated all round except at the very base, and seated on taper stalks rather less than half their own length; they are slightly wrinkled on the upper side, but very much so on the lower. The figs are seated on short stalks, have a pear-shaped figure, and seem to be as large as the fruit of the Sorb; usually they appear singly, but in some instances two have grown from the same axil. A deciduous shrub, capable of withstanding an ordinary winter, if planted in a dry situation. It was killed to the ground by the last severe winter. It grows freely in any good garden soil. It has no beauty as an object of cultivation, and is only interesting as a distinct half-hardy species.—*Journ. of Hort. Soc.*, vol. i.

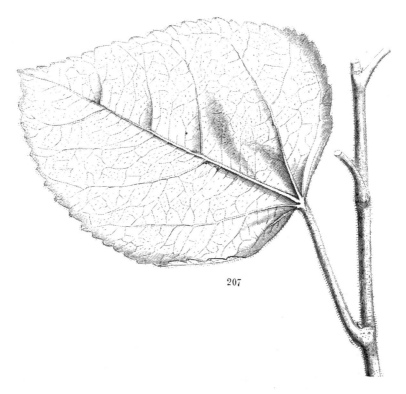

207

PLATE 64

L.Constans del. & zinc.

Printed by C.F. Cheffins, London.

[PLATE 64.]

THE AZURE PENTSTEMON.

(PENTSTEMON AZUREUS.)

———◆———

A hardy Herbaceous Plant from CALIFORNIA, *belonging to the Order of* LINARIADS.

═══════════════════

𝔖pecific 𝔠haracter.

THE AZURE PENTSTEMON.—Quite smooth, and somewhat glaucous. Leaves opposite, the lowermost oblong, stalked, the upper ones sessile, somewhat cordate, lanceolate, acute, occasionally split at the point, or incised at the edge. Racemes rod-like. Flower-stalks opposite, one-flowered, furnished with bracts. Sterile filament smooth.

PENTSTEMON *AZUREUS;* glaberrimus, glaucescens, foliis oppositis inferioribus petiolatis oblongis superioribus sessilibus subcordato-lanceolatis acutis nunc apice bifidis aut a latere incisis, racemis virgatis, pedunculis oppositis bracteatis unifloris, filamento sterili glabro.

Pentstemon azureus : *Bentham, Plantæ Hartwegianæ,* p. 327, no. 1879. *Journal of Horticultural Society,* vol. v., p. 144. *Our Volume* i., no. 105.

═══════════════════

THE remarks at No. 105 of our First Volume have prepared our readers for a knowledge of this handsome Perennial, which proves to deserve more attention at the hand of the gardener than was anticipated. It forms a fine tuft of slightly glaucous branches and leaves, producing in the autumnal months rods of bright azure blue flowers about 2 feet high.

Mr. Bentham compares it with *P. heterophyllus,* than which it is a very much finer species, readily distinguished by its blue flowers, growing singly and almost without stalks, in the axils of the upper leaves. It is also to be remarked that the foliage, although usually perfectly entire, exhibits every now and then a tendency to acquire divisions such as are represented in the accompanying Plate;

the leaves are sometimes split at the point into two sharp lobes, and sometimes become incised, or even serrulate. Their general tendency is, however, to be wholly undivided.

Like the other species of this genus from California, the Azure Pentstemon is readily multiplied by either seeds or cuttings; nor is it at all necessary to grow it in peat: on the contrary, it thrives perfectly in common garden soil. Since its introduction our winters have been so mild that we are unable to say with certainty that it is hardy; but there is no doubt that a hand-glass or a heap of leaves will give it all the protection it needs.

PLATE 65.

L. Constans del. & zinc.

Printed by C.F Cheffins, Lond.

[PLATE 65.]

THE LONG-LEAVED BROMELIA.

(BROMELIA LONGIFOLIA.)

———◆———

A Hothouse Perennial from GUIANA, *belonging to the Natural Order of* BROMELIADS.

Specific Character.

THE LONG-LEAVED BROMELIA. Leaves very long, scurfy, with spiny teeth, curved backwards, and extended into a long, linear, bristle-shaped point. Spike globose, nearly sessile, many-flowered. Bracts oblong, roundish, serrulate, with a sharp abrupt point, covered with white meal. Sepals linear-lanceolate, somewhat spiny, mealy, rather more than half as long as the petals.

BROMELIA LONGIFOLIA; foliis longissimis farinosis spinoso-dentatis recurvis in apicem longum linearem setaceo-acuminatum productis, spicâ globosâ subsessili multiflorâ, bracteis oblongis subrotundis serrulatis cuspidatis albo-furfuraceis, sepalis lineari-lanceolatis subspinosis furfuraceis petalis vix duplò brevioribus.

————————

Bromelia longifolia : *Rudge, Plantæ guianenses,* p. 31, t. 49.

————————

FOR this very fine Bromeliad we are indebted to **Mr.** Henderson of the Wellington Road Nursery, who exhibited it at the meetings of the Horticultural Society in August last, as the Tillandsia ————, of some manufacturer of Garden names. It is a true Bromelia, and was long since published in the work above quoted, with a figure in outline made from a dried specimen collected in Guiana by Martin.

Leaves from 1½ to 2 feet long, narrow, channelled, tapering to a fine point, coarsely spiny-toothed, white beneath, greyish green, and smooth on the upper side, gracefully curving away from the centre. Head of flowers like a rich rose-coloured cone, standing on a short stalk, with a few narrow crimson spiny bracts at its base, powdered with a white meal. The proper bracts are broadly ovate, concave, cuspidate, finely serrated, as long as the flowers. Ovary inferior, smooth, shining, sharply triangular, with six placentæ standing in pairs near the inner angles of three double partitions. Sepals keeled at the back, narrow, acuminate, slightly serrated, somewhat mealy. Petals not quite twice as long, erect, pink, obovate, apiculate, naked at the base. Stamens six, equal, as long as the petals. Style somewhat protruded, with three short slightly twisted stigmas.

The species is one of the prettiest of its race, which we are glad to perceive is gradually coming into favour among gardeners. For brilliancy of colour the Vegetable Kingdom hardly produces anything equal to that of many species of Bromeliads; witness the Vriesias, Æchmeas, Pitcairnias, and Billbergias already in cultivation.

PLATE 66.

L.Constans del. & zinc.

Printed by C.F.Cheffins, London.

[PLATE 66.]

THE SWEETEST AIR-PLANT.

(AËRIDES SUAVISSIMUM.)

————•————

A Hothouse Epiphyte from MALACCA, *belonging to the Natural Order of* ORCHIDS.

Specific Character.

THE SWEETEST AIR-PLANT. Raceme horizontal, oblong, many-flowered. Bracts dwarf, ovate, scarious. Sepals and petals oval, blunt, very much spreading. Lip horned, ascending, pressed to the column, three-lobed ; its lateral divisions oblong, somewhat toothed, the intermediate one dwarf, linear, bifid, quite entire.

AËRIDES *SUAVISSIMUM ;* racemo oblongo horizontali multifloro, bracteis nanis ovatis scariosis, sepalis petalisque ovalibus obtusis patentissimis, labello cornuto ascendente columnæ adpresso trilobo, laciniis lateralibus oblongis subdenticulatis intermediâ nanâ lineari bifidâ integerrimâ.

Aërides suavissimum : *Lindley in Journ. of Hort. Soc.,* vol. iv., p. 264. A. flavidum : *suprà,* vol. i., no. 372, a variety.

THIS plant has been introduced from the Straits of Malacca by Messrs. Loddiges, with whom it first produced its flowers in June 1849. At that time it was described as being in general appearance similar to *A. odoratum,* but in fragrance more balsamic and delicious. The sepals and petals were white, with a lilac tip ; the lip was pale nankeen colour, with a lilac streak along the centre of the middle lobe. It was said to differ from *A. odoratum* in the middle lobe of its lip being emarginate and much longer than the laterals, which are distinctly notched; and from *A. Quinquevulnera* in its not being at all serrated, as well as in its greater length.

Since that time we have had much finer specimens for examination from Mr. C. B. Warner, two of which are now figured, and we find that the flowers acquire a very distinct blush, instead of the paleness which was first described; the point of the spur is also a warm red, and the lip itself is a pale lemon-colour.

The great feature of this species is the small bifid middle lobe of the lip, which is sometimes longer than the lateral toothletted lobes, and sometimes much shorter, while the lateral lobes are quite entire. The latter form was called by us *A. flavidum* at No. 372 of our previous volume; the flowers were found to be glutinous and very fragrant; and of the lip the horn was green, the lobes pale yellow, the petals and sepals white dashed with pink.

At Plate 60 we gave some account of the species of Aërides, which constitute the first section of this genus, characterised by having a flat undivided lip, with an ear perhaps at the base. We now publish a continuation of that account.

AËRIDES §. *Labello trilobo; laciniá intermediá majore, v. lateralibus æquali.*

5. A. crispum *Lindl. in Wall. Cat.*, no. 7319; *Gen. and Sp. Orch.*, no. 6; *Bot. Reg.*, 1842, t. 55; A. Brookei *Bateman in Bot. Reg.*, 1841, misc. 116; foliis planis apice obtusis obliquis bilobis racemis multifloris nutantibus dupló brevioribus, sepalis petalisque subæqualibus obtusis, labelli maximi laciniâ intermediâ multò majore ovatâ retusâ serrulatâ basi bidentatâ lateralibus erectis acutis nanis, calcare cornuto incurvo.—*Peninsula of India.*—This charming species, in the magnitude of its blossoms, surpasses any other kind, the lip alone being upwards of an inch long, beautifully. tipped with rose-colour. The other parts of the flower are of brilliant white. In the form of the labellum the species slightly resembles *Aërides affine*. In its habit it is stiff and erect, the leaves being only five inches long, while the racemes are more than double that length.

6. A. falcatum *sp. nov.;* foliis coriaceis distichis obtusis apice obliquis mucronatis, racemis densis pendulis multifloris, labelli tripartiti laciniis lateralibus falcatis obtusis intermediâ obovatâ fissâ ciliatâ convexâ basi bicristatâ multò angustioribus et brevioribus, calcare brevi cum labello parallelo.—*East Indies.*—A very striking species, exhibited by Sir George Larpent at the June Garden Meeting of the Horticultural Society in 1847. The habit of the plant is that of *A. crispum;* the sepals and petals are white with a crimson speck at the point; the lip is crimson in the middle, white barred with rose at the edge, and on the lateral lobes.

7. A. cylindricum *Lindl. in Wall. Cat.*, no. 7317; *Wight's Figures*, t. 1744; ? Epidendrum subulatum *Retz. Obs.*, 6. 50; ? Limodorum subulatum *Willd. Sp.*, pl. 4, 126; foliis teretibus, racemis brevibus subbifloris, sepalis ovatis obtusis, petalis oblongis latioribus, labelli cucullati infundibularis laciniis lateralibus oblongis obtusis intermediæ carnosæ ovatæ obtusæ adnatis, calcare recto conico.—*Iyamally Hills, Coimbatore.*—" Leaves round, somewhat cylindrical. Racemes short, about two-flowered. Sepals ovate, obtuse. Petals oblong, broader. Lip cucullate, funnel-shaped; lateral lobes oblong, obtuse; adnate to the ovate, obtuse, fleshy, middle one. Spur straight, conical. Flowers white, or slightly tinged with red, lip reddish; middle lobe yellow at the base." (*Wight.*)

8. A. testaceum *Lindl. Gen. and Sp.*, no. 2; foliis loratis acutis bilobis inter lobos cuspidatis, racemis strictis simplicibus multifloris foliis longioribus, sepalis petalisque obovato-oblongis obtusis, labelli infundibularis laciniis lateralibus erectis obtusis intermediâ lineari patente apice dilatatâ reniformi bilobâ dentatâ lineis duabus elevatis callosis in disco, calcare conico incurvo.—*Ceylon, on trees.*—Peduncles spotted. Flowers the size of *A. Wightianum*, pale yellow, with a violet spot in the middle of the lip. Capsules clavate, six-angled.

9. A. Wightianum *Lindl. in Wall. Cat.*, no. 7320 ; *Wight's Figures*, t. 1669, under the name of *Vanda parviflora ;* foliis loratis apice obliquis obtusis bilobis inter lobos cuspidatis, racemis strictis simplicibus multifloris foliis longioribus, sepalis petalisque ovalibus anticis majoribus, labelli infundibularis laciniis lateralibus pedi columnæ adnatis obtusis intermediâ subcuneatâ apice trilobâ rotundatâ : disco lineis pluribus elevatis crispis cristato, calcare brevi conico.—*Iyamally Hills, Coimbatore.*—" Leaves strap-shaped, oblique at the base, obtuse, two-lobed, with a tooth between. Racemes straight, simple, many-flowered, longer than the leaves. Sepals and petals oval, the anterior ones larger. Lip funnel-shaped, lateral lobes adnate to the foot of the column, the middle one sub-cuneate, roundish, three-lobed at the apex ; disk crested, with several elevated crisp lines. Spur short, conical. Middle lobe of the lip deep lilac. Capsules club-shaped, six-angled. Flowers yellow." (*Wight.*)

AËRIDES §. *Labello trilobo ; laciniâ intermediâ multò minore (nanâ).*

10. A. odoratum *Lour. Fl. Cochinch.* 525 ; *R. Brown in Hort. Kew.* 5. 212 ; A. cornutum *Roxb. Mss. Bot. Reg.* t. 1485 ; foliis flaccidis apice obtusis obliquis, racemis pendulis multifloris foliis longioribus, labelli cucullati infundibularis laciniis lateralibus erectis cuneatis rotundatis intermediâ ovatâ acutâ inflexâ, calcare conico incurvo.—*Common in the hottest parts of India ; also in China and Cochin China.*—Flowers white, pink at the point, fleshy, very sweet-scented.

11. A. suavissimum (*aliàs* A. flavidum) *of this article.*

12. A. Quinquevulnera *Lindl., Sertum Orchidaceum,* t. xxx.; foliis ligulatis apice rotundatis obliquè emarginatis apiculo interjecto, racemis pendulis multifloris foliis longioribus, labelli cucullati infundibularis laciniis lateralibus erectis intermediâ oblongâ inflexâ denticulatâ, calcare conico incurvo.— *Philippines.*—Flowers slightly fragrant, speckled on white, with a purple stain at the end of each of its five divisions.

13. A. virens *Lindley in Botanical Register,* 1843, misc. 48, 1844, t. 41; foliis latis obliquè retusis, racemis pendulis multifloris, sepalis petalisque obovatis obtusis, labelli cornu acuminato ascendente lobis lateralibus apice denticulatis intermedio lanceolato medio canaliculato versus apicem denticulato.—*Java.*—" A beautiful species belonging to that set of Aërides of which *A. odoratum* was the first discovered. Like the flowers of that species, these are deliciously and very peculiarly sweet-scented, and not at all inferior in size. Each sepal and petal has a deep purple blotch at the end, while the remainder is a delicate soft French white. The lip is speckled with crimson, and bears in the middle an inflated, sanguine, serrated tongue. The leaves are much alike in all these plants, but here they are of a peculiarly bright green, which circumstance has suggested the name." According to Blume, his *A. suaveolens* (*Rumphia* III., t. 193, f. 1.) is this species ; but its small rose-coloured flowers are not such as we see in this country.

14. A. pallidum *Lindley, Gen. and Spec. Orch.,* no. 18 ; *Blume, Rumphia* III., t. 197., f. A. Dendrocolla pallida *Blume, Bijdragen* 290 ; "foliis carnosis rigidis canaliculatis oblique emarginatis inter lobos mucronatis, racemis ascendentibus foliis longioribus multifloris, perigonii phyllis obovato-oblongis obtusis, labelli cucullati lobis lateralibus retusis intermedio cuneato subtrilobo crispo, calcari conico incurvo.—(*Blume.*)"—*Timor, on trees ;* also *Philippines* (Cuming.)—Flowers showy, white. Lip with transverse pale rose-coloured bars. Spur yellowish at the tip. Flowers very sweet, like Lily of the Valley.

SPECIES TO BE EXCLUDED FROM THE GENUS.

A. tessellatum *Wight in Wall. Cat.*, no. 7318.   Epidendrum tessellatum *Roxb. Corom.* 1, t. 42.   Cymbidium tessellatum *Swartz. Nov. Act. Ups.* 6. 75; *Willd. Sp. Pl.* 4. 102.—This is a doubtful plant.   The specimens distributed by Dr. Wallich consisted of loose flowers of *Vanda Roxburghii,* and the leaves of some plant unknown.

A. appendiculatum *Wallich Cat.*, no. 7315.
A. tæniale *Lindl. Gen. and Spec.*, no. 7.   } probably represent so many genera.
A. difforme *Wallich, Sertum Orch., fronte* f. 7.

A. amplexicaule, and others, forming Blume's genus *DENDROCOLLA,* are plants concerning many of which we have at present no information.   With them must be associated the *Liparis? Prionotes* of the *Gen. and Species of Orchidaceous plants.*

GLEANINGS AND ORIGINAL MEMORANDA.

415. ABELIA UNIFLORA. *R. Brown.* A small hardy (?) evergreen shrub from the North of China, belonging to the Order of Caprifoils. Flowers whitish, produced in July. Introduced by Messrs. Standish and Noble. (Fig. 208.)

Our first knowledge of this shrub was derived from Mr. Reeves, who sent dried specimens from China about the year 1824. It was upon one of these, communicated to Dr. Brown, that the species was established in Dr. Wallich's *Plantæ asiaticæ rariores*, when the peculiarity of the solitary flowers, each subtended by three bracts (not eight, as is stated in De Candolle's Prodromus) was pointed out. The specimens in question had been collected in the province of Fokien, near Ngan-ke-hyen, in the Black Tea Country, lat. 25° N., long. 116° E. The plants in cultivation were procured by Mr. Fortune, and sent to Messrs. Standish and Noble, who consider the shrub to be hardy.

It forms a small erect, nearly smooth bush, with opposite or ternate leaves, ovate, and slightly toothed or entire, rather more coriaceous in our wild specimens than in those which we received from Bagshot. The flowers grow singly in the axils of the upper leaves ; their calyx usually consists of a pair of large obovate membranous sepals slightly toothed at the end ; occasionally three sepals are present. The corolla is rather longer, white, with a slight violet stain on the upper side ; it is remarkable for the great quantity of its spiral vessels, which are unusually tough ; there are four stamens, a filiform style, and a three-cornered stigma. The ovary, which is long, narrow, and rather pubescent, contains three cells, one of which contains a single pendulous ovule, while the other two are many-seeded ; at its base are three sharp triangular minute scales. It is probable that the *Abelia serrata* of *Siebold* and *Zuccarini,* is not distinct from this. In its hairiness it resembles the garden state of the plant now described, but the sepals in the specimen before us are shorter.

208

416. DAMMARA OBTUSA. *Lindley.* A greenhouse Coniferous tree, with very blunt oblong leaves. Native of the New Hebrides. Introduced by Mr. C. Moore. (Fig. 209.)

Of this remarkable species a plant has been received alive. It was found on the island of Aniteura, one of the new Hebrides, by Mr. Moore, who describes it as a tree similar in appearance to the Kauri of New Zealand (*Dammara Australis*), from which it is distinguished by the size and form of both leaves and cones. It grows to a great size, and produces a valuable timber, which is much used for ships' spars. The leaves are nearly four inches long by one and a quarter broad, very exactly oblong, with the end rounded off, without the least trace of point. The cone which I have received, and which seems to be full-grown, is three inches long by one and three quarters wide, somewhat cylindrical, with the ends rounded. The ends of the scales are convex, about four times as broad as long, and quite different in that respect from the spreading points of the New Zealand Kauri.—*Journ. of Hort. Soc.*, vol. vi.

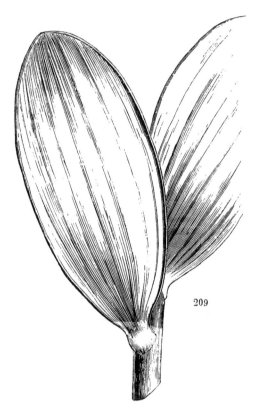

209

417. GEISSOIS RACEMOSA. *Labillardière.* A magnificent hothouse tree with long racemes of crimson flowers. Inhabits New Caledonia. Belongs to the Order of Cunoniads. Introduced by Mr. C. Moore.

This is, probably, the finest stove plant that has been introduced for several years. One plant has reached the Garden in good health. Mr. Moore describes it as " a native of the east coast of New Caledonia, in bare, exposed situations. Leaves woolly and slightly serrated when young, entire and glaucous when the plant arrives at a flowering state. It is a small tree, bearing the flowers, which are of a crimson colour, on the old wood in great abundance." The dried specimens sent home have opposite trifoliolate leaves of a firm leathery texture, with obovate, very obtuse leaflets, from 6 to 7 inches long, and between 3 and 4 inches broad. Between each pair of leaves is a sessile, amplexicaul, smooth, roundish, leathery stipule. The racemes of flowers are from 8 to 12 inches long, with stalks even longer than themselves, and bearing a pair or two, or an additional whorl, of great glaucous stipules like those belonging to the leaves. The flowers are rich crimson, packed closely like a Combretum, with globular buds, 4 leathery ovate sepals, shaggy with hairs in the inside, and 8 stamens with crimson filaments nearly an inch long. When in flower these must produce a gorgeous effect, at least equal to that of *Combretum grandiflorum.* From the above slight description, the botanical reader will see that this plant does not quite agree with Labillardière's figure and description ; but I am unable to say that Mr. Moore's is a distinct species of *Geissois* without the opportunity, which I do not possess, of instituting a comparison with authentic specimens.—*Journ. of Hort. Soc.*, vol. vi.

418. ACER CIRCINATUM. *Pursh.* A most beautiful hardy deciduous tree from Oregon, with purple and white flowers, and leaves rich crimson in the autumn. Introduced by the Horticultural Society. (Fig. 210.)

There is probably no hardy tree in this country more eminently beautiful than this, if tree it can be called, for it seems rather a bush. In the spring, when its leaves unfold, they are preceded by long crimson leaf-scales, from two to four to each twig ; the leaves when they first come are thin, semitransparent, and a clear light green ; at the same time peep out little tufts of purple flowers, with white petals ; and in the autumn the plant seems on fire with the rich red of the foliage, more rose-coloured, and not less intense, than that of the most scarlet of Oaks.

Sir William Hooker tells us that the species is found wild on the Great Rapids of the Columbia River, and is common along the north-west coast of North America, between lat. 43° and 49°. Mr. Douglas observes that it is exclusively confined to the woody mountainous country that skirts the shores, and there, among the pine forests, it forms almost impenetrable thickets. The branches are pendulous and crooked, often taking root, as is the case with many species of the genus Ficus. Bark smooth, green when young, white when fully grown. The wood is fine, white, and close-grained, very tough, and susceptible of a good polish. From the slender branches of this tree the native tribes make the hoops of their *scoop-nets,* which are employed for taking salmon at the Rapids, and in the contracted parts of the river. It is said to form a tree twenty to forty feet high.

419. ACER VILLOSUM.   *Wallich.*   A noble tree, from the Himalayahs, with the aspect of a Sycamore.   Introduced by Messrs. Osborne & Co., of the Fulham Nursery.   (Fig. 211.)

Dr. Wallich tells us that this is a very large tree, inhabiting the high alps of India, approaching towards those of perpetual snow in Sirmore and Kamaon, ripening its fruit in November, at which time " the very fragrant flowers also begin to appear." Dr. Royle says it is only " seen with Pines and Birches on the loftiest mountains, which are for many months covered with snow." In its general appearance this may be compared to the common Sycamore, but is a much finer looking tree, its leaves being thicker, greener, and larger ; besides which they are covered with a close fur on the underside, although smooth above ; in the autumn they assume a peculiar nankeen tint. The plants in the possession of Messrs. Osborne & Co. have not yet blossomed ; but our Herbarium tells us that the " fragrant " flowers come out in close panicles, covered with long yellowish hairs. Undoubtedly this is one of the finest hardy deciduous trees yet introduced.

It is to be hoped that India will soon yield us her other alpine Sycamores, of which there are three, viz., 1. *A. sterculiaceum* Wallich, found near the summit of Mount Sheopore, and very like *A. villosum*, except that it is nearly destitute of hairs. The trunk of this is said to be three feet in diameter, and the flowers white.—2. *A. caudatum* Wallich, so called because the palmate leaves have the lobes extended into tails. In this the leaves are scarcely more than three-lobed, and are sharply and doubly serrated. Dr. Wallich says it is a native of the highest regions of Nepal, towards Gossain Than, as also of Sirmore and Kamaon. Dr. Royle found it growing in company with *A. villosum*. It is a remarkable and handsome species.—3. *A. cultratum* Wallich, the leaves of which are heart-shaped, and deeply divided into seven much acuminate *undivided* lobes, besides being much smaller and thinner than in the two preceding species. It is " a larger tree, native of the regions towards the Himalayah, in Kamaon and Srinaghur." Dr. Royle who also found it says, that its wood is " white, light, and fine-grained." Dr. Wallich suggests its being allied to the *Acer pictum* of Japan, to which we must add that it is little different from Bunge's *Acer truncatum* from Northern China.

420. CARAGANA TRIFLORA.   *Lindley.*   A hardy half-evergreen shrub from Nepal.   Flowers yellow, in May.   Belongs to the Leguminous Order.   Introduced by the East India Company.   (Fig. 212.)

212

Under the name of *C. triflora* this Indian shrub has been for some years in cultivation, having been raised in the garden of the Horticultural Society, from Nepal seeds presented by the Court of Directors of the East India Company. It forms a dwarf nearly smooth bush, the leaves being pubescent only when young. The leaflets are in four or five pairs, oblong, blunt or obovate, and placed upon a petiole which is spiny at the point like the stipules. The peduncle is not more than half the length of the leaves, erect, slender, smooth, and bears at its end an umbel of from two to three flowers (seldom more than three), whose pedicels are more than twice as long as the setaceous weak bracts at their base. The calyx is very slightly downy ; at the base of its tube is a pair of membranous mucronate bracts ; its teeth are subulate, soft, and much shorter than the tube. The petals are yellow, but not so deep as in Chamlagu. The bush inhabits the Himalayas, where, with other spiny Caraganas, it forms a vegetation analogous to that of our Furze and Whins. It has been described in the *Botanical Register* for 1845, misc. 56.

421. BEGONIA (DIPLOCLINIUM) SEMPERFLORENS.   *Link & Otto.*   A succulent hothouse perennial from Brazil.   Flowers white in winter and spring.   Belongs to the Order of Begoniads.   (Fig. 213.)

Although not one of the handsomest, this is one of the most useful of its genus in consequence of the long time during which it remains in flower. It is altogether green in the herbage, and white in the flowers, and destitute of hairiness. Its brittle tapering stems are from one and a half to two feet high. The stipules are long, membranous, and obtuse. The leaves are succulent, roundish ovate, slightly and setaceously serrate, becoming nearly entire when old ; at the base they are very slightly oblique, and not at all heart-shaped. The flowers appear in little terminal and axillary

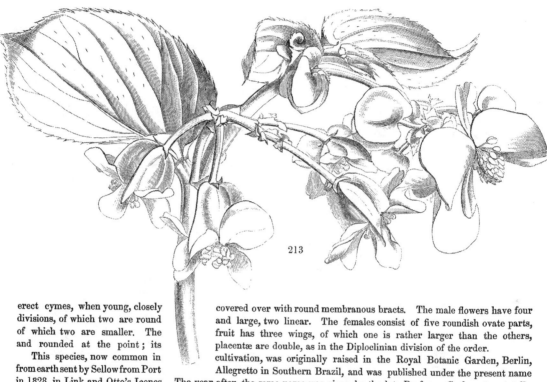

213

erect cymes, when young, closely divisions, of which two are round of which two are smaller. The and rounded at the point ; its

This species, now common in from earth sent by Sellow from Port in 1828, in Link and Otto's Icones.

covered over with round membranous bracts. The male flowers have four and large, two linear. The females consist of five roundish ovate parts, fruit has three wings, of which one is rather larger than the others, placentæ are double, as in the Diploclinian division of the order. cultivation, was originally raised in the Royal Botanic Garden, Berlin, Allegretto in Southern Brazil, and was published under the present name The year after, the same name was given by the late Professor Graham to a totally different species, with red stems and flowers, and leafy persistent stipules, figured in the *Botanical Magazine*, t. 2920, and perhaps not sufficiently distinct from *B. spathulata*.

422. PENTSTEMON WRIGHTII. *Hooker.* A beautiful perennial, with glaucous leaves and loose panicles of crimson flowers. Native of Texas. Flowers in June and July. Introduced at the Royal Botanic Garden, Kew.

This is a charming new *Pentstemon*, very distinct from any hitherto known to us, and which will prove a great acquisition to our gardens. It was discovered by Dr. Wright in Texas, and has been distributed among the very interesting dried collections of that gentleman, without any name, by Dr. Engelmann. Root perennial (?) Stem erect, including the panicle a foot and a half or two feet high, terete, branching from the base, and there rather woody, purplish-brown and scarred from the fallen leaves, the rest glaucous, and bearing distant pairs of opposite very glaucous leaves, few in number, spathulate, that is oblong or obovate, entire, tapering into a stalk, all except the uppermost pair at the base of the panicle, which are ovate, oblong, quite sessile, truncated or even cordate at the base. From above these the elongated panicle arises, a foot or more long, bearing several pairs of small ovate bracteas, from the axil of each of which is seen a two-flowered peduncle, with a small ovate bracteole at the base of each pedicel. Flower drooping. Calyx with minute, glandular hairs, shortly campanulate, the five acute entire segments spreading. Corolla deep rich rose-colour, slightly downy, the tube about an inch long, ventricose on the underside towards the mouth. Limb an inch broad, spreading horizontally, cut to the base into five nearly equal rotundate lobes. Stamens included. Filaments quite glabrous, flexuose. It appears to grow and flower freely, but we are not yet certain whether it is quite hardy. Like other species of the genus, it will probably be found to succeed best if a stock be kept in pots under a frame in winter, and planted out in the open ground in spring. It is increased by seeds, which it produces readily.—*Bot. Mag.*, t. 4601.

423. PHILESIA BUXIFOLIA. *Lamarck.* A half-hardy shrub from Chiloe and Patagonia, with stiff deep green leaves and rich crimson tubular flowers. Belongs to Philesiads. Blossoms in September. Introduced by Messrs. Veitch and Co.

Among evergreen non-coniferous shrubs, this is probably the finest which Messrs. Veitch have imported, even although it should require a greenhouse. Dr. Hooker enumerates it "among the handsomest plants of the Antarctic American Flora ; occurring along the coast, from the Strait of Magalhaens to Valdivia." Mr. Lobb writes of it thus :—" The Philesia is a plant of very slow growth. In its native country it forms large masses on trunks of trees and rocks, throwing out long slender stems, which creep along beneath the decayed bark, and over rocks that are partly covered with soil. The roots, which proceed from the internodes of the stem, are few and brittle, and very difficult to preserve. No plant that I have seen requires so much care in moving." In another place he writes :—" It is a splendid thing, and probably the most valuable plant of my collections. It often covers trunks of trees and rocks. Sometimes it grows erect, but when found in that state it seldom exceeds a foot in height, and is always growing about the base of dwarf stunted wood, similar to coppice in England. The flowers are produced near the extremity of the branches, have a campanulate form, and are sometimes not less in size than the common Tulip, of a deep rose colour. The petals are thicker in substance than any other flower that I have seen. I have traced it from the level of the sea to the snow line, and it flowers more freely at a great elevation."—*Journ. of Hort. Soc.*, vol. vi.

424. PHYSOCHLAINA GRANDIFLORA. *Hooker.* A hardy perennial with pale green flowers strongly marked with darker veins. It belongs to Nightshades, is nearly allied to Henbane, and is not handsomer. Native of Thibet.

Our garden is indebted for the seeds of this plant to Lieutenant Strachey, who gathered them on the plains of Thibet, at an elevation of 15,000 feet above the level of the sea. Root probably perennial. Stem herbaceous, a good deal branched, terete, clothed everywhere, as well as the foliage, with glandular down. Leaves alternate, petiolate, ovate, acute, penninerved, thrice as long as the petiole. Panicle terminating the branches, leafy. Pedicels elongated. Floral leaves gradually passing into bracteas. Flowers drooping. Calyx shortly campanulate, sharply five-toothed, in fruit much enlarged and elongated, becoming tubular or cylindrical, and then erect. Corolla more than an inch long, slightly curved downwards, between campanulate and infundibuli-form, the mouth spreading, the lobes short, rounded, obtuse ; the colour is yellow-green with a slight tinge of purple, marked with longitudinal purple lines connected by oblique transverse ones. A strong-rooted, hardy, herbaceous plant, thriving in any kind of garden-soil. It may be increased by dividing the roots, which should be done in autumn or early in the spring.—*Bot. Mag.*, t. 4600.

214

425. LINDLEYA MESPILOIDES. *Humboldt & Kunth.* A fine, sweet-scented, evergreen, half-hardy bush, from Mexico. Flowers white. Belongs to the Rosaceous Order. Introduced by the Horticultural Society. (Fig. 214.)

This plant is an evergreen tree, of small size, looking very much like *Mespilus grandiflora*, but with flowers as sweet-scented as the Hawthorn bloom. It belongs to a small set of Rosaceous plants, of which one, the *Kageneckia cratægifolia*, is occasionally seen in this country. The late Professor Don attempted to distinguish them as a peculiar natural order, but unwisely, and on erroneous grounds. That they are really nothing more than Rosaceous plants, is proved by this plant grafting readily on the common Thorn and the larger kinds of Cotoneaster, in which way it is propagated. But although *Lindleya* and its allies are by no means to be separated from *Rosaceæ*, they form a peculiar group, remarkable for their capsular fruit and winged seeds, the latter a circumstance not hitherto observed in other plants of the order.

The botanical peculiarity of the present genus consists in its carpels joining together at the very base into a solid pistil, although their upper halves, as well as the styles, are entirely distinct. And so, in like manner, when the fruit is ripe, it becomes a hard capsule, the thick bony lobes of which separate freely at the upper half, but not at the lower without violence. In our gardens the plant proves to be about as hardy as an *Escallonia*, but not more so. It remains in flower for a month or six weeks after the beginning of July. In its native country it forms an evergreen slender-growing shrub, twelve to fifteen feet high, near the natural bridge called Puente de Dios, 45 miles N.E. of Real del Monte, at an elevation of 6500 feet above the sea. It also occurs sparingly near the Hacienda de Santa Ana, in the state of Oaxaca, always preferring a dry chalky soil.—*Bot. Register*, vol. xxx., t. 27.

426. ALLAMANDA NERIIFOLIA, *of Gardens.* A fine-looking stove plant with large yellow flowers, the origin of which is unknown. Belongs to Dogbanes. Introduced by Lucombe and Co. Flowers in June.

Its habit is extremely different from that of any de-scribed species, as is the form of the corolla, with its singularly short contracted base of the tube, swollen and angled at the base, and the very elongated upper portion: the colour is a deep almost golden yellow, and it is streaked with orange. " The plant, from which the speci-men was cut, is now only three feet high. It commenced flowering when but eighteen inches high. The first and largest cluster consisted of thirty finely expanded flowers. An evergreen shrub, with copious and handsome foliage everywhere glabrous. Leaves oblong, on short petioles, acuminated, deep green above, pale and reticulated beneath. Panicle of many flowers, in reality terminal, but, by and by, lateral from innovations or young shoots which again terminate with clusters of flowers. Calyx of five, ovato-lanceolate, spreading lobes. Corolla smaller than in *A. Schottii* or *A. Aubletii*, but deeper-coloured than either, and elegantly streaked with orange. In shape it is quite different from both, the lower and con-tracted portion of the tube being very short, swollen, and angled at the base, the rest of the tube or faux is bent at an angle and much elongated, between funnel-shaped and campanulate : the lobes are rounded, acute, spreading. Stamens and pistils quite included.—*Bot. Mag.*, t. 4594.

427. EPIDENDRUM VOLUTUM. A hothouse epiphyte from Central America. Flowers greenish white. Introduced by G. U. Skinner, Esq., in 1849. (Fig. 215.)

Epidendrum (Osmophytum) *volutum ;* caule tereti membranaceo-vaginato apice 3-phyllo, foliis lineari-ob-longis pergameneis inæqualibus, racemo sessili stricto flexuoso bracteis magnis glumaceis distantibus pedunculis longioribus, sepalis petalisque linearibus revolutis æqua-libus acutis, labello subrotundo-ovato subcordato decurvo leviter trilobo cuspidato callis 2 ad basin plicisque tribus obsoletis.

This very distinct species of Epidendrum flowered last summer in the garden of the Horticultural Society, where it was received from Mr. Skinner. Like many of the Osmophytes, as the section of Epidendrum to which it belongs is called, the stem is merely terete, and not swollen into a pseudobulb. The flowers are greenish white, and of no beauty. It is very easily known by the great alternate paleaceous bracts, planted on a somewhat zigzag rachis, and reaching nearly to the middle of the foot-stalk and ovary of each flower.

215

428. BRASAVOLA ACAULIS. A singular epiphyte from Central America. Belonging to Orchids. Flowers cream-colour, at midsummer. Introduced by G. U. Skinner, Esq. (Fig. 216.).

B. acaulis ; foliis teretibus rectis et flore subsessilibus, sepalis petalisque linearibus patulis æqualibus, labelli laminâ subrotundo-ovatâ ungue cucullato duplo longiore.

This singular plant approaches *B. glauca* in its manner of growth, the stem being so short as to be scarcely perceptible, and *B. grandiflora* in the size and form of the lip. It is strikingly different from all others. Only one flower appears on a very short stalk, greenish white, with some tendency to spotting ; the firm and narrow sepals, about three inches long, and curving round the lip, the flat part roundish ovate, and about twice as long as the rolled up claw. The leaves are remarkably short and stiff. The only plant we know of is in the garden of the Horticultural Society, where it was received from Mr. Skinner, and was reported to have pink flowers.

216

PLATE 67.

L.Constans del. & zinc.

Printed by C F Cheffins, London

[PLATE 67.]

THE LONG-FLOWERED CENTRANTH.

(CENTRANTHUS MACROSIPHON.)

———•———

A Hardy Annual from the SOUTH OF SPAIN, *belonging to the Natural Order of* VALERIANWORTS.

Specific Character.

THE LONG-FLOWERED CENTRANTH.—A smooth annual. Stem erect, dwarf, branched, stout, fistular, glaucous. Leaves ovate ; the lower short-stalked, entire, or obscurely toothed, obtuse ; the upper sessile, sharply toothed, more or less deeply cut at the base, with linear lobes. Panicles dichotomously corymbose, compact. Bracts narrow, linear, with a membranous edge. Flowers deep rose-colour. Tube of the corolla three times as long as the fruit ; spur thrice as short as the fruit. Pappus with black (?) feathery plumes, united at the base by a membrane.

CENTRANTHUS *MACROSIPHON;* annuus, glaber ; caule erecto pumilo ramoso crasso fistuloso glaucescente ; foliis ovatis, inferioribus breviter petiolatis integris aut obsolete dentatis obtusis, superioribus sessilibus acute dentatis, basi magis minusve profunde incisis, laciniis linearibus ; paniculis ad ramorum et caulis apicem dichotomè corymbosis densifloris ; bracteis angustè linearibus membranaceo-marginatis ; floribus intensè roseis ; corollæ tubo fructu triplo longiori, calcare fructu triplo breviore ; pappi setis nigris (?) plumosis infernè membranâ inter se coalitis. *Walpers.*

Centranthus macrosiphon : " *Boissier Diagnos. pl. nov. orient.* III. 57," according to *Walpers' Repertorium,* vi. 80.

ACCORDING to M. Boissier this grows on damp house-tops (*tecta humida*), in the warmest western part of the kingdom of Grenada, near the town of Estepona. This author distinguishes it from *C. Calcitrapa* by its upper leaves being less cut, and only at the base; by its flowers being four times as long, and deep rose-coloured ; by its spur being thrice as long ; and by the part of the corolla placed between the calyx and base of the spur being extremely short, not, as in *C. Calcitrapa,* half as long as the fruit. According to the specific character given by M. Boissier, it should also have black pappus, of which we see no sign.

It was introduced to this country by the Horticultural Society, who received it from Messrs. Vilmorin, of Paris. We find it to be an excellent autumnal annual, with masses of rich ruby flowers of various tints, giving the heads a sparkling varied appearance, which our colourers are unable to imitate.

PLATE 68.

[PLATE 68.]

THE BLAND AMARYLLIS.

(AMARYLLIS BLANDA.)

———◆———

A Stove Bulbous Plant from the CAPE OF GOOD HOPE, *belonging to the Natural Order of* AMARYLLIDS.

Specific Character.

THE BLAND AMARYLLIS.—Flowers horizontal, closely umbelled, with a short tube. In *A. Belladonna* the flowers are somewhat erect and fewer, and there is no tube at all.

AMARYLLIS *BLANDA;* floribus horizontalibus densè umbellatis. In *A. Belladonna* flores pauciores suberecti, et tubus nullus.

Amaryllis blanda : *Ker in Botanical Magazine,* vol. xxxv. t. 1450 ; *Herbert, Amaryllidaceæ,* p. 277.

FOR the opportunity of figuring this beautiful plant we are indebted to Mrs. Bellenden Ker, in whose collection, at Cheshunt, it flowered last September. It was bought in a lot marked "Hybrid," at the sale of the plants of the late Dean of Manchester. Kept in a stove it grew very rapidly, soon going to rest; and suspicion arose that the stove was not the right place for it. But last year, while apparently at rest, it threw up two large flower-stems, loaded with fragrant bloom. The bulb is covered with a pale brown soft skin, composed of multitudes of thin layers filled with cottony threads. The leaves are grass-green, an inch and a quarter broad, with a regularly-rounded point. Twelve or thirteen beautiful large flowers, thin, delicate French white, changing to pink, load the end of the scape, forming an umbel of great sweetness.

There can be no doubt that it is the identical *Amaryllis blanda* figured thirty-eight years ago in the " Botanical Magazine," and now almost unknown in cultivation, concerning which the late learned Dean of Manchester makes the following remarks :—

"This beautiful plant was found by Niven, who collected for Mr. Hibbert, and I believe has never since been met with by any collector.   I purchased one of the bulbs when Mr. Hibbert disposed of his collection, and Mr. Griffin had another.   Mr. Knight, of the King's Road, Chelsea, who had the rest, killed them by planting them in the open ground, which they will not endure in this country, and I believe there are no bulbs of it in Europe but the produce of those two.   I lost two by planting them in front of the stove; one died the first winter, the other only lingered till the second.   The leaves of this and the following species, when cut by frost or drought at the points, will not continue to grow like those of *Belladonna*.   It requires an airy situation in the greenhouse in winter, drought and dry heat in summer, and will then flower magnificently in September. Whatever may have been the growth of its leaves, it will not flower if it is left in a cold situation while dry."

It is very near the well-known *Belladonna Lily*, especially a pallid variety of that species, not rare in gardens; but it is quite different in constitution, and clearly distinguished by its flowers having a very perceptible tube, instead of rising abruptly from the top of the ovary.   The flowers are moreover more numerous, more fragrant, and more horizontal.   The late Mr. Ker, when he originally published it, observed that it would be superfluous to particularize differences, which a comparison of the figures and descriptions of the two plants would so easily show.   " In *Belladonna* the segments of the corolla do not cohere at all beyond their base, but converge in such way as to give the appearance of their so doing; the leaves are of a dark dingy green, scarcely more than half an inch broad, and never attain a length in any way equalling the scape; which circumstances are here mentioned, because they were omitted in our account of that species.   *Blanda* is a native of the Cape of Good Hope, where it was gathered by Sir Joseph Banks; was sent to Miller in 1754 by Van Royen from Holland, and flowered in the Chelsea Garden."

PLATE 6

I.Constans del. & zinc.

Printed by C.F.Cheffins.

[PLATE 69.]

THE SHOWY GRAMMATOPHYL.

(GRAMMATOPHYLLUM SPECIOSUM.)

———◆———

A Stove Epiphyte, from the MALAY ARCHIPELAGO, *belonging to the Order of* ORCHIDS.

Specific Character.

THE SHOWY GRAMMATOPHYL. Caulescent. Leaves in two rows, sword-shaped, nerveless. Scape erect racemose. Flowers coriaceous, as long as their stalks. Sepals and petals obovate, oblong, wavy, obtuse. Middle lobe of the lip velvety, with three smooth ribs reaching higher than the middle, with lines of hairs next the ribs in the bottom.

GRAMMATOPHYLLUM *SPECIOSUM ;* caulescens, foliis distichis ensiformibus enerviis, scapo erecto racemoso, floribus coriaceis pedicellis æqualibus, sepalis petalisque obovato-oblongis undulatis obtusis, labelli lobo medio velutino ultra medium glabro abruptè tricostato circa costam in fundo piloso.

Grammatophyllum speciosum : *Blume, Bijdragen,* p. 377, tab. xx. ; *Lindl. Gen. and Sp. Orch.* p. 173 ; *Blume, Rumphia,* vol. iv., p. 47, t. 191 ; *Museum Botanicum Lugduno-Batavum,* i. 47.

At last is realised the long cherished wish to see this in flower. After years of patience, Mr. Loddiges succeeded in persuading it to expand a few blossoms last summer, all of which were in a monstrous state except one. Nevertheless they enabled our artist to prepare the accompanying figure which gives some idea of what the plant is ; only the flowering scape proceeded from the top instead of the bottom of the stem, whence it arises if in a natural condition.

According to Blume this noble plant inhabits Java, and other islands of the Indian Archipelago, as well as Cochinchina, where it was found at Pulo Dinding on trees by Mr. Finlayson. "The vigour of its growth, and the extraordinary size of its flowers, render it the Queen of Orchids." The mountaineers of Java call it *Kadaka sousourou*. We have it from the Straits of Malacca, where the late Mr. Griffith found it, with "a scape six feet high, and an inch in diameter at the base."

Dr. Blume says, that, in its wild state, the stems are from two to three feet high, straightish, cylindrical, from an inch to an inch and half in diameter, covered at the base with rudimentary leaves only, but towards the top closely loaded with foliage. The perfect leaves are in two rows, equitant at the base, very much spreading or curved backwards, from one to two feet long and an inch wide, striated, shining, smooth, with a central rib channelled on the upper side, and bluntly protuberant on the lower. The flowering stem, or scape, springs directly from the base of the leaf-stem, is from five to six feet high, and bears a profusion of blossoms for about half its length. Each flower stands upon a curved stalk, about three inches long, and has the same or a greater diameter; in texture it is between fleshy and leathery; outside it is pale lemon-colour, inside a brighter yellow, marked with numerous roundish brown spots, arranged with much irregularity; all the parts are somewhat oblong, a little narrowed at the base, the two lateral sepals having a slightly curved figure, as in *Renanthera*. The lip is equally three-lobed, rolled round the column, and about half the length of the sepals; it is attached by a moveable joint to a pouch at the base of the column; the lateral lobes are acute and smooth, except near the middle line of the lip, where they are hairy; the middle lobe is ovate, more coriaceous, somewhat longer, and covered with a thick felt, except in the very middle, where three raised naked lines extend to within a short distance of the tip.

As has been already stated, the specimen that flowered with Mr. Loddiges produced but one perfect flower. All the others were in various deformed states, of which the following, whose columns are represented in the annexed cut, were the most important.

No. 1. Sepals 2 and petals 2, decussating. Column opposite one of the sepals, with a hooked spur proceeding from the lower edge of the flat narrow face of the column, and curving upwards;

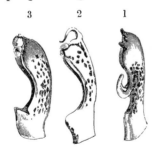

lower half of column terete. Stigma a perforation. Pollen 2 globular masses, united at the base, and excavated behind. Gland 0. No. 2. The same, except that the sepals were broader; the two petals directed forward, and not exactly decussating; no spur on the column; a long cirrhus growing from the hinge of the anther. No. 3. Sepals 3, the two lateral partially united at the base. Petals in natural state. Column excavated at the base, and prominent there, with two teeth at the upper edge of the projection. It was apparently as perfect as in the natural condition; but no lip formed upon it.

These changes may contribute hereafter to our knowledge of the theoretical value of the parts of the flower of Orchids.

The genus Grammatophyllum was originally proposed by Dr. Blume, and was adopted in the Genera and Species of Orchidaceous plants by one of us, in consequence of its having a pair of pollen masses hollowed out behind, and seated each upon one end of a horse-shoe shaped gland; by which circumstance alone it appeared to differ from Cymbidium. Since that time the learned Dutch botanist has published a fine figure and a detailed account of the plant, from which we beg to translate the following passages :—

"As a genus this differs from *Cymbidium* in having the column elevated in front into a hump, and in the peculiar insertion of the pollen masses into the extremities of a horse-shoe shaped caudicle. A more exact examination will show whether *Cymbidium giganteum* of Wallich also belongs to it. The *Cymbidium elegans* of Lindley is the type of a new genus, which is quite distinct both from Cymbidium and Grammatophyllum in its long club-shaped column and two pear-shaped pollen masses furrowed at the back, disjoined, and fixed transversely to a common flat oval caudicle. This may be named CYPERORCHIS *elegans*. Equally different from Grammatophyllum is my genus LEOPARDANTHUS, remarkable for its short broad obliquely truncate column, to the base of which the saccate lip is adnate, and its bifid caudicle peltate and recurved in front, on the two legs of which are seated elliptical pollen masses furrowed at the back. This *Leopardanthus scandens* is a terrestrial caulescent plant, climbing up the trunks of trees, with distichous sword-shaped ribbed sheathing leaves, axillary, erect, many-flowered scapes, and middle-sized spotted flowers."—*Rumphia*, iv., p. 47. Concerning the Cyperorchis above named, we shall have something to say in an early number of this work. Leopardanthus is unknown to us, except by a figure in Blume's *Museum*.

The following is the state of our acquaintance with the species belonging to this genus.

1. GRAMMATOPHYLLUM SPECIOSUM ; the subject of the foregoing remarks.

Lip and pollen masses of Grammatophyllum.

2. GRAMMATOPHYLLUM FASTUOSUM ; foliis . . , scapo erecto subcorymboso, floribus coriaceis pedicellis 2-plo brevioribus, sepalis petalisque obovatis undulatis obtusis anticis incurvis, labelli lobo medio velutino infra basin glabro 3-costato, circa costam in fundo scabro-piloso.

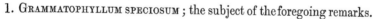

A single specimen of this majestic plant was sent us from Malacca by Griffith, who found it on trees, and saw no leaves. It is upon the whole a finer species than *Gr. speciosum* itself. The flowers are not quite so large, but in consequence of the far greater length of their stalks (as much as 5 inches), they form a kind of corymb. Their colour is unknown to us. Both sepals and petals are more spathulate, the 2 anterior of the former far more incurved, and the naked ribs of the lip, instead of reaching almost as far as the point of its middle lobe, and then terminating abruptly, gradually lose themselves in the pile of velvet at the very base of the lobe. The lip appears moreover to be much more coriaceous.

3. GRAMMATOPHYLLUM SCRIPTUM *Blume, Rumphia*, vol. iv., p. 48; *Museum Botanicum*, vol. i., p. 47.

Under this name is now arranged Rumphius's *Angræcum scriptum*, and the Cymbidium, Epidendrum, or Gabertia founded upon it. We have never seen any specimens ; the plant is not in cultivation, nor do we find that Blume himself has any personal knowledge of it, for what he says of it is chiefly, if not wholly derived from Rumphius. That author describes two sorts of *Angræcum scriptum*, one of which grows upon the Mango tree, the other on the Calappa Palm tree. The FIRST has great flattish conical pseudobulbs (*bursæ*), deeply furrowed lengthwise, and each bearing 3 or 4 long narrow leaves like those of "Helleborus albus or Hyris," thick, firm, narrow below, widening upwards, above a foot long, 3 digits wide, and having in the middle 3 ribs, which do not project much. The flower-stem has no leaves, is 4 or 5 feet high, a little curved at the top, from which the flowers grow regularly one above another as in Hyacinths. The flowers themselves are as big as a Narcissus, composed of 5 outer petals, narrow at the bottom, broader at the top, some yellow, others greenish-yellow and inscribed with large spots and characters like Hebrew letters, but not distinct ; they are reddish-brown, and different in different flowers. In the middle is a rolled up petal resembling a goblet, of a paler colour, streaked with brown or purple lines. Of this he says there are certain varieties—The SECOND, which grows on the Calappa Palm tree, and of which he gives a figure. Its bursæ or pseudobulbs are as in the other, but smooth, not jointed, and they are not so much furrowed lengthwise. The leaves are shorter, broader, and thicker, without any ribs except that in the middle, from 13 to 16 inches long, and 4 broad ; they grow from the young pseudobulbs (*ex teneris bursis*). The flowering stem is from 5 to 5½ feet long, as thick as the little finger, and covered with bloom for two-thirds of its length. The flowers are not unlike those of the first sort ; they have 5 green petals, one of which is always turned inwards. These petals are painted with thick brown characters, which form no particular figure, but are chiefly spots ; among them however you may make out the letters A. I. O. The lip is paler than the other petals and marked lengthwise with brown lines.

In Ternate, according to Rumphius, the matrons, and especially the wives, sisters and daughters of the Kings (who are all called *Putri* in Malay, and *Boki* in the Moluccas), so entirely appropriate these flowers to themselves that a common woman and especially a slave would offer them a great affront if she were to put them on her head. The flowers are reserved exclusively for the great ladies, who cause them to be sought for in the forests, and braid them in their hair, saying that nature herself has shown that such flowers are not fit for people of low degree, since they grow nowhere except in high places ; hence they are called *Bonga boki* and *Bonga putri*, or the Princesses' flower. It would also appear that the Malay gentlemen make from the seeds a philtre which has a surprising effect upon the ladies who swallow it.

It is far from improbable that these two sorts of lettered Angrec are distinct species, judging from what is said of their leaves. The second kind may indeed be identical with the species next to be mentioned.

4. GRAMMATOPHYLLUM multiflorum *Lindley in Botanical Register*, 1838, misc. 80, 1839, t. 65 ; *var.* tigrinum *Ib.* 1842, t. 69 ; foliis lanceolatis coriaceis subundulatis striatis in apicem pseudobulbi ovati insidentibus, scapo incurvo, racemo erecto longissimo cernuo multifloro, floribus membranaceis, sepalis oblongis obtusiusculis planis, petalis acutis subconformibus angustioribus, labelli trilobi pubescentis medio hirsuti lobo intermedio plano oblongo rotundato lateralibus erectis subfalcatis, jugo in medio carnoso elevato ad basin lobi intermedii interrupto in costas 2 lamelliformes producto, columnæ margine supra basin elevato flexuoso incurvo foveam altam obconicam circumdante.

The two varieties of this species, one with whole-coloured, the other with spotted flowers, were brought from the Philippine Islands, and differ altogether from both *Gr. speciosum* and *fastuosum* in having thin membranous flowers, instead of the thick leathery texture of those two species, and drooping not erect racemes. Dr. Blume suggests that this plant may not be different from his *Gr. scriptum* ; and it is not impossible that it may include Rumphius's *second* sort of *Angræcum scriptum* (the one he has figured) ; but it can scarcely be compared with his first sort, either in leaves, or pseudobulbs, or in the markings of the flower, which have nothing grammical about them or resembling Hebrew characters. The name, then, given by us in 1838 will not have to be disturbed.

* Grammatophyllum ? pulcrum *Spanoghe in Linnæa* xv., 477, a Timor plant, is known only by name.

** Grammatophyllum ? Finlaysonianum *Lindley, Genera and Species of Orch.* p. 173, is now called *Bromheadia palustris*.

GLEANINGS AND ORIGINAL MEMORANDA.

429. SALVIA CANDELA-BRUM. *Boissier*. A stately hardy perennial, from the south of Spain, with large leafless panicles of violet and yellow flowers. Belongs to Labiates. Introduced from Malaga by E. Delius, Esq. (Fig. 217.)

This is one of the many curious plants discovered in the south of Spain by the indefatigable zeal of M. Boissier, the distinguished Swiss traveller. Not having within reach the work in which it was published, we can only state that it is an inhabitant of the Sierra Nevada of Grenada, in the district, we presume, which produces the *Abies Pinsapo*. Its leaves are like those of the common sage, to which it is nearly related ; but it throws up a glaucous branching naked panicle, three feet long, of large flowers, of which a fragment is represented in the annexed cut. These flowers have a greenish-yellow upper lip, and a rich violet lower one ; and they would produce a fine appearance if any considerable number opened at the same time. It happens however that they are short-lived, and drop off soon after expansion, so that no seeds are ripened, and the plant has a shabby appearance. Probably the flowers would hold on, and the beauty of the species be much enhanced, if it were grown in a little bottom heat. In its native country, where everything is favourable to its growth—hot dry weather when in flower, and warm damp weather while growing—it is reported to be a noble-looking thing, even in the rich gardens of Grenada.

217

430. IMPATIENS PLATYPETALA. *Lindley.* (*aliàs* I. pulcherrima *Dalzell.*) A handsome tender annual, from Tropical India. Flowers large, violet-purple. Belongs to the Order of Balsams. Introduced some years since by Messrs. Veitch.

This very handsome stove plant is not uncommon in gardens, to which it was introduced some years since by Messrs. Veitch. It was well figured in the *Botanical Register* for 1846, under the name of *I. platypetala*; and has lately been admirably represented in the *Botanical Magazine* (t. 4615) under the name of *I. pulcherrima*, which must be cancelled. Sir William Hooker there speaks of it to the following effect :—

" One of the finest of the Indian Balsams. Mr. Dalzell found the plant near Warree, in the Southern Concon, Bombay, and seeds were sent to us in 1850. The plants continued to bear flowers during most of the summer months. Like the other tropical species of *Impatiens*, a succulent, tender annual. The seeds should be sown in spring, and if placed in a gentle heat they will soon vegetate. When the young plants are of sufficient strength, they must be potted singly in small pots, and duly shifted into larger ones as they increase in size, which they will do rapidly if supplied with rich soil and plenty of water, and kept in a close pit or frame. A few may be planted in the open air in a sheltered place ; but they are liable to suffer from too free an exposure to the winds and rain of this climate."—[To this we may add, that, when regarded as a stove annual, this species merits universal cultivation. It flowers all winter long.]

431. PODOCARPUS NUBIGENA. *Lindley.* A beautiful hardy evergreen bush, or tree. Native of Southern Chile. Belongs to Taxads. Introduced by Messrs. Veitch & Co. (Fig. 218.)

218

P. nubigena ; (Eupodocarpus) monoica foliis linearibus mucronatis subtus glaucis, pedunculis solitariis receptaculo oblique bilobo obovato brevioribus, fructibus oblongis oblique obtuse apiculatis.

This is one of the " Yews " mentioned by Mr. Lobb under Saxe-Gothæa (our No. xix., p. 112), and in general aspect it sufficiently justifies the name. It is a plant with stiff, linear, deep-green leaves, having a broad double glaucous band on the underside. The male flowers are unknown. The fruit is drupaceous, and grows singly in the axils of the leaves on very short stalks ; the receptacle is obovate, and obliquely two-lobed ; the nut oblong, slightly bossed, and curved inwards at the point. No species of Podocarp yet discovered agrees with this. *P. Lamberti*, from Brazil, has leaves green on both sides, and globose fruit. *P. chilina* has broader leaves, also not glaucous, and fruit with very long stalks. *P. andina* has the fruit in spikes. Messrs. Veitch possess only two small plants of this species. —*Journ. of Hort. Soc.*, vol. vi.

432. LEUCOTHOE NERIIFOLIA. *De Candolle.* (*aliàs* Andromeda neriifolia *Schlechtendahl ; aliàs* Agarista neriifolia *Don ; aliàs* Leucothöe crassifolia *De Candolle ; aliàs* Andromeda crassifolia *Pohl ; aliàs* Agarista Pohlii *Don.*) A handsome greenhouse evergreen shrub, with panicles of crimson flowers. Native of Brazil. Belongs to Heathworts (*Ericaceæ*), near Andromeda.

This handsome plant quite corresponds with what we believe to be *L. neriifolia* De Cand., first found by Sellow in tropical Brazil, then by Mr. Gardner in Minas Geraes. It is worthy of a place in every greenhouse. Our flowering specimen was communicated by Mr. Cunningham of Comeley Bank

Nursery, without any history of its introduction. The ovary is remarkable for producing at its base, in all the flowers we examined, simple or branched subulate filaments, which from their position may be considered abortive stamens. A moderate-sized shrub, with very coriaceous, evergreen, oblong leaves, gradually acuminated at the point and then ending in a mucro, the base cordate, footstalk very short, glabrous on both sides, minutely reticulated beneath. Raceme solitary, from the upper axils of the leaves, much longer than they, nearly erect, very handsome. Rachis and pedicels red, indistinctly rugulose (under a glass) with very minute acicular bracteoles. Calyx red, deeply five-cleft. Corolla bright scarlet, between ovate and urceolate, very thick and fleshy : limb moderately large, of five acute spreading lobes. The species thrives in light peat soil well drained. It should be placed in a cool shady house or pit, especially in summer, for, like the generality of Ericaceous plants from elevated regions, it is apt to suffer by full exposure to the sun of this climate.—*Bot. Mag.*, t. 4593.

433. TROPÆOLUM SPECIOSUM. *Endlicher & Pöppig.* A hardy climbing perennial, with brilliant scarlet flowers, native of Chiloe and Patagonia. Blossoms all the summer. Belongs to the Order of Indian Cresses (*Tropæolaceæ*). Introduced by Messrs. Veitch in 1847.

Among the garden treasures imported from the temperate parts of South America, this is one of the most valuable and least known. It scrambles up sticks or bushes to the height of five or six feet, and bears an enormous quantity of scarlet flowers among a tender pale green foliage. In the winter it dies down to the perennial roots. It has been well figured in the Botanical Magazine, t. 4323, but the colour there does injustice to the plant, which, at a distance, looks like a mass of the scarlet cloth from which soldiers' jackets are made. The first discoverer of it was Mr. Pöppig, who found it in the subandine regions of southern Chile. Mr. Lobb says that it inhabits "*cool shady places*, often covering the branches of shrubs, and displaying a profusion of dark crimson velvety flowers." The words in italics are, no doubt, the key to the cultivation of the species. It can scarcely be said to be much known in cultivation, although introduced for four years. No doubt it has been generally lost ; perhaps, as in our own case, by giving it a warm sunny border. The experience of Messrs. Veitch shows that it cannot bear direct sunshine, or exist in a soil subject to dryness. With them it runs about in an American border, under a north wall, where the noon-day sun never reaches ; its creeping roots force their way through the neighbouring gravel walk, and the strong vigorous shoots form so compact a mass of flowers and leaves, that the wall seems as if lined with scarlet cloth. There is no question that the plant is perfectly hardy, if the border where it grows is rather damp, and if a few leaves are used to shelter the roots in winter.

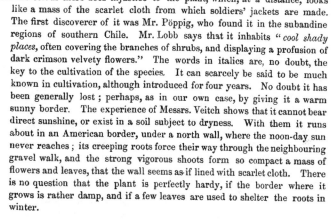

434. FAGOPYRUM CYMOSUM. *Meisner.* (*aliàs* Polygonum emarginatum *Wallich ; aliàs* P. acutatum *Lehmann ; aliàs* P. cymosum *Treviranus.*) A hardy perennial, with white sweet-scented flowers, from Nepal. Blossoms in autumn. Belongs to the Order of Buckwheats. Introduced by the East India Company. (Fig 219.)

This is a fine-looking perennial creeping-rooted plant, with stout erect stems 3 feet high. The leaves are triangular and tapering to each angle, with a cordate base, large and flat, forming an excellent relief to large spreading cymose panicles of pure white flowers, resembling those of the common Buckwheat (*Fagopyrum esculentum*), and like them succeeded by triangular fruit with winged angles. Flowering in the autumn, sweet-scented, and quite hardy, this species is particularly well suited to decorate shrubberies, and places where plants of a stouter

219

growth are wanted than those of parterres. It is a good bee plant, yielding abundance of honey. Any kind of garden soil appears to suit it. There seems to be little difference between it and the *Fagopyrum triangulare*, except that in the latter the branches of the inflorescence are usually in pairs, longer, and more divaricating, while the fruit is said to have two of its angles blunt, a circumstance we have not had the opportunity of verifying. Prof. Meisner remarks that the hollow stem of this plant is a circumstance without parallel among Polygonums, but he was not then acquainted with *Fagopyrum triangulare*.

435. GAULTHERIA NUMMULARIÆ. *De Candolle.* (*aliàs* G. nummularioides *D. Don*; *aliàs* G. repens *Blume.*) A trailing evergreen greenhouse plant, with white flowers, and reddish purple berries. Native of the Himalayas. Belongs to Heathworts. Raised in the garden of Her Majesty at Frogmore. (Fig. 220.)

This pretty little evergreen trailer was raised by Mr. Ingram in the Royal Gardens at Frogmore, whence only we have received it. Naturally it inhabits alpine places in India, from Gossain Than, and Nepal, to Java, for Blume's *Gaultheria repens* does not appear to be different. Dr. Royle, who has figured the plant in his *Illustrations of the Botany of the Himalayan Mountains*, t. 63, says that it occurs on Gossain Than, and is the only species found by him in the more northern portion of the Himalayan Mountains. Mr. Lobb gathered it on the Khasija Hills, and sent it to Messrs. Veitch. Griffith seems to have seen it on the Bhotan Mountains, near Tassyassy, "on wet banks." Probably it is a greenhouse plant. The stems are not thicker than pack-thread, are covered with brown hairs, and trail upon the ground, forming a close entangled carpet. The leaves are sometimes nearly circular, whence its name, or they acquire an ovate form, and are pointed ; at their edges, and all over the underside, are scattered the same kind of stiff

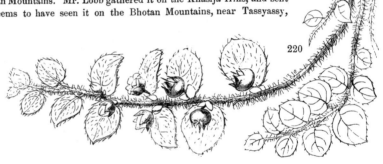

220

brown hairs as clothe the stem, (in order to show these, the accompanying figure represents the underside chiefly; the upper side is smooth.) The small white flowers grow singly in the axils of the leaves, and are entirely hidden by them. They are succeeded by reddish-purple glabrous fruit, growing on very short stalks, hidden by two or three smooth brown scarious cucullate bracts. The breadth of the leaves in our wild specimens varies from one quarter to three quarters of an inch. We may be censured for taking De Candolle's specific name *Nummulariæ*, instead of the older one of *num-mu-la-ri-o-i-des*, but we prefer the former to such a barbarously constructed uncouth name as the last.

436. SAXIFRAGA FLAGELLARIS. *Willdenow.* (*aliàs* S. aspera *Bieberstein*; *aliàs* S. setigera *Pursh.*) A hardy Arctic perennial, with golden yellow flowers. Introduced at Kew.

Not one of the many expeditions that have gone out to discover a "north-west passage," or in search of the many brave and excellent officers and men of the Erebus and Terror whose fate is yet unknown to us, but has prosecuted researches in various branches of natural history—botany in particular. The flora of the Arctic regions, consequently, is as well known as that of any portion of civilised Europe. Living plants from those regions are always desiderata, for our climate, especially in the latitude and in the proximity of London, is very unsuited to their preservation, and they soon perish. A box filled with various growing plants has been collected at Cornwallis Island, and sent to the Royal Gardens of Kew, by Capt. N. Penny, commanding the ship Albert, in conjunction with his very intelligent medical officer, Dr. Sutherland, and among them this curious and rare Saxifrage in a *flowering* state. It is drawn and lithographed and now published in little more than a month from its being landed in England, in October 1851. The present species of *Saxifraga* inhabits the Caucasian and Altaic Alps, as well as the rocky mountains of North America in about lat. 42°, to Melville Island in the extreme north and Behring's Straits to the west. Closely allied species are found in the Himalaya. It has received the appropriate name of the Spider plant from the sailors of our Arctic Expeditions. This diminutive plant will, we fear, like most Arctic plants, not last long in cultivation, owing to the impracticability of placing it under conditions of climate similar to those of its native countries. It there remains, for about ten months of the year, in a dormant state, buried under snow ; on the melting of which it springs immediately into growth, and, being stimulated by the warmth and continuous light of the sun during the short Arctic summer, comes rapidly to maturity, producing flowers and multiplying by means of viviparous stolons. During this short period the soil is thawed to a depth of from eighteen inches to two feet, the earth below remaining in a frozen state throughout the year, showing that vegetable life in the Arctic regions is entirely dependent upon solar influence. Such being the circumstances amidst which this plant lives, it should be kept in a state of rest during winter, which, under the influence of our varying temperature, is difficult ; for even if this and other Arctic plants are placed, in winter, in what we call a cool temperature, we still find them in a growing state, by which they become weak and soon exhaust themselves.—*Bot. Mag.*, t. 4621.

437. LOMATIA FERRUGINEA. *R. Brown.* A half-hardy shrub from South Chile, with beautiful ferruginous foliage. Belongs to Proteads. Introduced by Messrs. Veitch & Co.

This charming plant is growing in the open air in the nursery at Exeter, but we fear that it will not generally bear the climate of this country. It, however, deserves a place wherever beautiful foliage is valued. According to Cavanilles, it forms a shrub ten to twelve feet high, with ferruginous branches. The leaves are deep green, bipinnatifid, ferruginous when young, from six to twelve inches long, with some of the leaflets occasionally lobed. The flowers, which have not been produced in England, appear in short erect racemes from the axil of the leaves, and are said to be green outside, and crimson inside. Mr. Lobb does not say where it grows naturally, but according to Cavanilles, it inhabits S. Carlos in Chile, in places occasionally overflowed by salt water.

438. CALLICARPA JAPONICA. *Thunberg.* A half-hardy deciduous shrub, with small pinkish flowers. Native of Japan. Blossoms in August and September. Belongs to the Order of Verbenes. Introduced by Dr. Siebold. (Fig. 221.)

This is an inelegant soft-wooded shrub, growing two or three feet high, and having the branches closely covered when young with a short fur composed of stellate hairs. The leaves are stalked, about three or four inches long, serrated

221

except at the two ends which are entire, when full-grown destitute of hairs on the upper side, but downy on the veins of the underside ; in addition to this the under surface is sprinkled with pale yellow glands, not discoverable without a magnifying glass ; in form the leaves vary from oblong, tapering to each end, to almost a rhomboidal outline. The flowers grow in dense axillary racemes, which are many-flowered, nearly smooth, and a little longer than the leafstalks. The calyx is cup-shaped, very obscurely four-toothed, or altogether truncate. The corolla is pale pink, nearly equally divided into four blunt lobes, three times as long as the calyx, with four projecting stamens. The plant has not beauty enough to be worth growing in choice collections.

Siebold and Zuccarini have pointed out (*Floræ Japonicæ familiæ naturales*, part 2, p. 30) the error committed by

M. Schauer in referring this species to *C. longifolia*, a still less attractive plant, figured in the *Botanical Register*, t. 864, and now apparently lost in gardens. The form of the leaves is quite different, as are their serratures, which in fact are apt to disappear altogether in *C. longifolia*, whose cymes of flowers are smaller, with more conspicuous teeth to a firm and fleshy, not as in this case thin and membranous, calyx. *C. longifolia* is a southern plant, much more tender than this, which we believe occurs exclusively in Japan, whence we have wild specimens from Zuccarini, differing only in a looser and longer inflorescence and larger leaves.

439. FAGUS OBLIQUA. *Mirbel.* A fine evergreen tree from Southern Chile. Belongs to Mastworts. Introduced by Messrs. Veitch & Co.

This is, probably, a hardy evergreen tree. Mr. Lobb says :—" It inhabits the slopes of the Andes, from the level of the sea, to the line of perpetual snow It in general attains the height of forty to fifty feet, with a stem as straight and as smooth as the Pine." According to Captain King, as quoted in Hooker's " Flora Antarctica," this sort of Beech tree grows to a considerable size. The plant in cultivation grows freely in the open air at Exeter, and has a graceful appearance. In some respects the foliage is more like that of a Hornbeam than a Beech. The leaves are between lozenge-shaped and lanceolate, serrated, with strong straight veins, and are of a beautiful pale green colour.—*Journ. of Hort. Soc.*, vol. vi.

440. CAMPTOSEMA RUBICUNDUM. *Hooker & Arnott.* (*aliàs* Kennedya splendens *of Gardens*, and *Meisner's Plantæ Preissianæ*, 1. 89 *in notâ*.) A beautiful greenhouse twiner, of the Leguminous Order, from South Brazil. Flowers scarlet.

A very handsome climber, long ago described from

222

dried specimens in the *Botanical Miscellany*, and for some time cultivated in Germany, and since in England as *Kennedya splendens*. It was so named, as we learn from Mr. Bentham, by Meisner, who cautiously observes, " Originis ignotæ ; " while Dr. Walpers confidently says, " Hab. in Nova Hollandia." It has the habit of a New Holland *Kennedya*, but it is a native of southern Brazil and the adjacent Argentine provinces. It is only lately that, being trained immediately under the glass of the Palm-stove, it has yielded flowers with us. The racemes remind one of those of *Laburnum* or of *Wistaria sinensis*, but they are of a deep ruby-red colour. A climbing shrub of great length ; the older portions of the stem as thick as one's finger, and reticulated, as it were, with pits or hollows in the oblong areoles. Young leafy branches slender, terete, herbaceous, glabrous. Leaves distant, on long petioles, trifoliolate; leaflets petiolulate, oblong, or oblong-elliptical, retuse, glabrous, glaucous beneath. Racemes on rather long peduncles, compound, eight to ten inches in length, drooping, many-flowered. Calyx with two small bracteas at the base, tubular-campanulate, somewhat two-lipped, and irregularly four to six-lobed. Petals of the corolla deep ruby-red, nearly equal. Vexillum partially reflexed, ovate, clawed, with two blunt teeth at the base of the lamina. Alæ and carina oblong, clawed, each petal with a blunt tooth at the base of the lamina. Stamens diadelphous (9 and 1). Ovary linear, on a long stipes, and tapering into a subulate style. Legumen three inches long, stipitate, compressed, downy, acute. A stove-climber, well adapted for training up rafters or on trellis-work, and which grows freely, especially if planted in a bed of good rich soil. Where there is not sufficient room for it to extend, it may be treated as a pot-plant, and trained upon a trellis fixed to the pot ; but we have not found it, either way, to flower very readily. It may be increased by cuttings, placed in heat under a bell-glass.—*Bot. Mag.*, t. 4608.

This species is not very uncommon in Gardens, and was long since figured under its garden name in *Paxton's Magazine of Botany*. Its magnificent flowers would ensure its universal cultivation if the plant could but produce them. From the preceding remarks it would seem to require more light and heat than it usually receives.

441. STENOCARPUS FORSTERI. *R. Brown.* An

evergreen greenhouse shrub, of little beauty, from New Caledonia. Introduced by the Horticultural Society. Belongs to Proteads. (Fig. 222.)

Of this a live plant has been received from Mr. Moore, who speaks of it as a small Proteaceous plant, not uncommon on the east coast of New Caledonia. It is a bush with obovate, retuse, flat, veinless, or slightly three-ribbed leaves tapering to the base, and umbels of small apparently white flowers. It will not prove of any horticultural interest.— *Journ. of Hort. Soc.*, vol. vi.

442. SALPIGLOSSIS SINUATA. *Ruiz and Pavon;* flava. (VARIETIES OF COLOUR :—1. S. atropurpurea *Graham;* 2. S. straminea *Hooker;* 3. S. picta *Sweet;* 4. S. Barclayana *Sweet.*) A handsome hardy annual, from Chili, with flowers of various colours, deep purple, straw-colour, variegated, parti-coloured and bright yellow. Belongs to Linariads. (Fig. 223.)

We have little doubt that Mr. Bentham is quite right in reducing to one species the many coloured forms of Salpiglossis that our gardens contain, for neither in

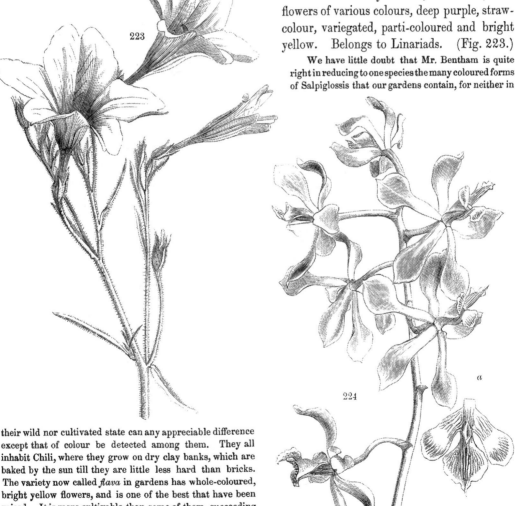

their wild nor cultivated state can any appreciable difference except that of colour be detected among them. They all inhabit Chili, where they grow on dry clay banks, which are baked by the sun till they are little less hard than bricks. The variety now called *flava* in gardens has whole-coloured, bright yellow flowers, and is one of the best that have been raised. It is more cultivable than some of them, succeeding perfectly well if sown in beds in the open air, not allowed to be crowded, and treated in all respects like a Lobel's Catchfly, a Collinsia, or any such well-known plant.

443. EPIDENDRUM REPLICATUM. An orchidaceous epiphyte, with yellowish flowers stained with brown, and a white and pink lip. Native of New Grenada. Introduced by Sigismond Rucker, Esq. (Fig. 224; *a*, a magnified representation of the lip flattened.)

E. replicatum ; (Encyclia hymenochila acuta) floribus dense racemosis, sepalis oblongo-lanceolatis acutis, petalis rotundatis unguiculatis apiculatis, labelli trilobi lobis lateralibus oblongis subtruncatis apice reflexis intermedio longiore crispo rhombeo acuminato lateribus omnino replicatis.

A very pretty species, exhibited by Mr. Rucker at the last July meeting of the Horticultural Society, when it received a silver Knightian medal. We presume it to be one of Mr. Linden's importations. The racemes were closely many-flowered, about 18 inches high. The sepals and petals are dull yellow, stained with brown in the middle below the point, but with a yellow border all round ; the former are oblong-lanceolate and acute, the latter are linear below, and then spread out into a circular disc, terminated abruptly by a small point. The lip, which is white, streaked with pink, is remarkable for the manner in which the two sides are turned downwards, so that their backs actually touch. Most nearly allied to *Epidendrum diotum,* a native of the same country.

444. ONCIDIUM SCHLIMII. *Linden.* A tall rambling Orchidaceous epiphyte, from Central America. Flowers yellow and brown in November. Introduced by Mr. Linden.

O. Schlimii ; (Plurituberculata) foliis binis rectis ensiformibus, scapo subscandente racemoso-paniculato racemis flexuosis, labello bilobo rotundato basi angustiore auriculato cristæ tuberculis quatèr ternis, columnæ alis triangularibus utrinque acuminatissimis.

We received flowers of this plant last November from the Fence near Macclesfield, where it had produced a scape five feet long. It had been purchased by Mr. Brocklehurst at one of Mr. Linden's sales. From dried specimens in our Herbarium, collected for Mr. Linden, we learn that it had been found by Funck and Schlim on the 7th of October, 1846, in the province of Merida, at the height of 7000 feet above the sea. The pseudobulbs are narrow, and bear a pair of sword-shaped thin leaves, from six to nine inches long. The panicle is wavy, weak, inclined to scramble, and bears, at intervals of about two and a half inches each, short racemes or imperfect panicles, not longer than the intervals themselves. The flowers are rather smaller than in *Oncidium reflexum,* near which the species will stand ; they are bright yellow, slightly and irregularly barred with brown. The peculiar form of the wings of the column—triangular, with the two free ends much acuminated and standing higher than the anther itself—renders it easy to identify the species. We cannot find that Mr. Linden's name, which we adopt, has been anywhere published up to this time.

445. CATASETUM SANGUINEUM. (*aliàs* Myanthus sanguineus *Linden.*) A terrestrial Orchid, from Central America, with greenish flowers, speckled with brown or dull red. Blossoms in October and November. (Fig. 225.)

C. sanguineum ; (Myanthus) sepalis petalisque oblongis acutis secundis, labello carnoso subrotundo rostrato serrato et lacero basi fimbriato foveâ altâ triangulari in medio.

This plant is not uncommon in collections under the name of *Myanthus sanguineus,* by which it has been dispersed at Mr. Linden's sales. We received the flower which furnished the annexed figure from Thomas Brocklehurst, Esq., of the Fence. Mr. William Pass, the gardener there, describes it as a strong-growing species with pseudobulbs six or seven inches long, and light glaucous green leaves. The flowers are in a close raceme, not at all handsome, notwithstanding the name, for the blood-red spots are quite dimmed by the dull green ground on which they are placed. This plant differs from *Catasetum saccatum* in having much smaller flowers, with the sepals and petals all turned upwards, the lip more lacerated than fringed, except quite at the base, and the opening of the pouch triangular without ribs instead of being crescent-shaped with very conspicuous elevations on the side next the base.

225

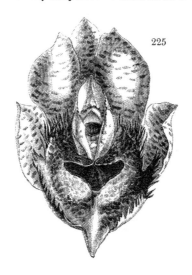

PLATE 70.

I. Constans del. & zinc.

Printed by C.F.Cheffins, L

[PLATE 70.]

THE LONG-CALYXED CHINESE AZALEA.

(AZALEA INDICA, *CALYCINA*.)

◆

A Greenhouse Shrub, from CHINA, *belonging to the Order of* HEATHWORTS.

Specific Character.

THE *LONG-CALYXED* CHINESE AZALEA. A noble variety, obtained from China, with very large deep rose-coloured spotted flowers ; the segments of the calyx leafy, and as long as the tube of the corolla.

AZALEA INDICA, *CALYCINA ;* varietas insignis e Chinâ allata, floribus maximis intensè roseis maculatis, calycis laciniis subfoliaceis tubo corollæ æqualibus.

THIS striking plant flowered last year in the garden of the Horticultural Society. It had been sent from China by Mr. Fortune, and put aside as some unimportant variety allied to *A. indica, phœnicea.* Its blossoms show it to be a much finer thing, especially remarkable for the very large size of the corolla, which, when fully expanded, measures more than three inches and a half in diameter, and is of a rich rose-colour, strongly speckled with bright crimson spots. There is also a peculiarity in its calyx, which is so long and leafy as to touch the re-entering angles of the corolla. We cannot, however, imagine that this circumstance gives our plant any claim to the rank of a species. On the contrary, it would rather seem to warn botanists against placing too high a value upon circumstances which, however striking, are merely exceptional and fugitive.

In some respects this may be compared to the plant called *Rhododendron Championæ,* in the *Botanical Magazine,* t. 4609; and which, although common among rocks in a ravine at Port Victoria, Hong Kong, is as yet unknown in gardens. That plant, however, is represented as being far more hairy, with great glutinous bracts, and its flowers are not spotted. We avail ourselves of the present opportunity to republish Sir William Hooker's description of *Rhododendron* (Azalea) *Championæ,* together with Captain Champion's notes upon four other Azaleas seen by him on Hong Kong.

1. *Rhododendron* (Azalea) *Championæ*. A shrub nearly seven feet high; branches terete, dichotomous; younger ones clothed with long, spreading, glandular bristles. Leaves much confined to the apex of the branches, shortly petioled (petiole glandular-bristly), lanceolate, shortly acuminate, reticulated, plane at the margin, dark green above, rather rusty-coloured beneath, the margin and costa and veins and veinlets clothed beneath and rough with short, harsh, bristly hairs. Flower-buds at first enclosed in a strobilus of large, imbricated, very glutinous, deciduous bracteas. Umbels four to six-flowered. Peduncles hispid with glandular hairs. Calyx, especially the margins, equally hispid, deeply cleft to the base into four erect, almost linear-subulate, rather long segments or sepals. Corolla four inches across, tube rather short, campanulate, white. Limb four inches across, deeply cut into five obovate-oblong, obtuse, unequal-veined segments, the upper one the broadest: the ground colour in our figure is white, the lobes, especially the apex and margins, are tinged with delicate rose-colour. But there is another state of the flower described by Captain Champion as the more usual colour, "delicate white, the upper lip pale yellow towards the centre, and copiously dotted with ochre." Stamens ten. Filaments much protruded, slightly curved upwards. Style equalling the stamens in length. Stigma a depressed disc. Capsule five to six-celled, elongated, nearly two inches long and three lines wide, cylindrical, straight, clothed with glandular bristles, "dehiscing from the base upwards, but remaining attached to the central axis."

2. *Azalea indica*, *phœnicea*. Is of common occurrence in Hong Kong in ravines. It flowers early in spring, and towards March appears in great beauty about waterfalls, by the side of streams, and on rocks or mountains, especially towards the eastern side of the island.

3. *Azalea squamata*. One of Mr. Fortune's species, produces a few flowers early in winter, but bursts into luxuriant blossom when the fogs and humid atmosphere about February and March have set in. Its lilac blossoms in mass look well at a distance, but the shrub, being then nearly destitute of leaves, has not on near approach the gay appearance which the scarlet-flowered *A. indica* presents.

4. *Azalea ovata*. This, which had been previously described by Mr. Fortune, from more northern China, grows, I believe, on the Black Mountains of Hong Kong; it there flowers in March with *A. indica* and *A. squamata*. It has almost rotate flowers, white with dark purple specks on the centre and adjoining lobes.

5. *Azalea myrtifolia*. A new species, quite distinct from *A. ovata*. A shrub from four to five feet high, much branched; twigs longer than in *A. squamata*, and shorter than in *A. indica*, quite smooth, cinereous, and striated with silver or pink-brown. Leaves alternate, crowded towards the extremities of the branchlets, short-petioled, from ovate to oblong or slightly rhomboid (largest one inch long by six lines broad), usually slightly emarginate at apex, with the midrib often prolonged into an acumen, quite smooth, bright green above, glaucous or pale beneath, and grossly reticulately veined. Flowers terminal, solitary or in pairs, from an elongate, ovate whorl of yellowish, or slightly glutinous, permanent scales; these scales ovate, smooth. Flowers in bud campanulate. Corolla, when expanded, from one inch two lines to an inch and a half in diameter, almost rotate, and cleft to near the base. Segments five, oblong, two upper slightly largest, pure white, the three lower with dark violet specks. Stamens five. Filaments hairy. Anthers opening by terminal pores. Style long, curved. Stigma clavate and ten-lobed at the apex. Calyx and pedicel pinkish, glutinous, puberulous, the former small. Capsule five-celled, above three lines in length, globosely ovate. Grows on the Black Mountain, Hong Kong, on rocks with *A. squamata* and *A. indica;* first seen, March, 1849, by Lieut.-Col. Eyre, of the Royal Artillery.

L. Constans, del. & zinc.

Printed by C.F. Cheffins, Lond.

[PLATE 71.]

THE GENTIAN-BLUE PENTSTEMON.

(PENTSTEMON GENTIANOIDES.)

———————◆———————

A Hardy Herbaceous Plant, from MEXICO, *belonging to the Order of* LINARIADS.

═══════════════

Specific Character.

THE GENTIAN-BLUE PENTSTEMON. Erect and tall. Leaves lanceolate, the uppermost widely stem-clasping, acuminate, and smooth. Panicle long, somewhat interrupted, leafy at the base. Flower-stalks short, béaring more blossoms than one. Segments of the calyx broadly ovate, acute, scarcely membranous. Tube of the corolla widely bell-shaped. Sterile filament, smooth, very blunt.

PENTSTEMON *GENTIANOIDES ;* erectus, elatus, foliis lanceolatis superioribus latè amplexicaulibus acuminatis glabris, paniculâ elongatâ sub-interruptâ basi foliatâ, pedunculis plurifloris abbreviatis, calycis segmentis lato-ovatis acutis vix membranaceis, corollæ tubo amplè campanulato, filamento sterili glabro apice retuso.— *Bentham.*

———————————

Pentstemon gentianoides : *Bentham in De Cand. Prodromus,* x. 323 *; aliàs* Chelone gentianoides, *Humboldt, Bonpland, and Kunth, Nov. Gen. and Sp.,* ii. 364, t. 172.

═══════════════

Mʀ. BENTHAM has determined, by the examination of authentic specimens, that this is the plant to which Professor Kunth applied the term *Gentianoides,* and not the long-flowered crimson kind so named in the *Botanical Register* and *Botanical Magazine,* and now everywhere in gardens. That kind, having been found near the Real del Monte mines by the collector Hartweg, is in future to be called *P. Hartwegii,* and is readily distinguished by its long narrow flowers, growing in a loose naked panicle, not in a long leafy raceme, as in this instance.

The fine species now represented is as hardy and easily managed as *P. Hartwegii* itself. Its flowers are short, inflated, very distinctly bell-shaped, and bright azure blue; but their effect is greatly impaired by the numerous floral leaves among which they are mixed.   Humboldt and Bonpland found it in Mexico, in cold places, on the slope of the snow-capped mountain of Toluco, at the height of 10,500 feet above the sea, flowering in September.   Hartweg sent it to the Horticultural Society from one of his stations named Anganguco, where it grew in pine-forests.

We suspect that the beauty of the plant would be materially enhanced if it were grown in soil that would check its excessive vigour.   If, instead of being four feet high, it could be dwarfed to eighteen inches, or two feet, it would be a lovely bedding-out species.

PLATE 72.

I. Constans del & zinc.

Printed by C.F. Cheffins. London.

[PLATE 72.]

THE PINK BUTTERFLY PLANT.

(PHALÆNOPSIS ROSEA.)

◆

A most beautiful pink-flowered Epiphyte, from MANILLA, *belonging to* ORCHIDS.

Specific Character.

THE PINK BUTTERFLY PLANT. Leaves oblong, leathery, sharp and recurved at the point (from eight to ten inches long). Flowers twelve or thirteen, about an inch in diameter, at the end of a stiff, ascending, drooping, branched, lateral peduncle (eighteen inches long). Sepals spreading, oblong-lanceolate, rather acute, equal, white slightly tinged with pink. Lip ascending, deep violet, with the lateral segments linear-spathulate, oblique, incurved, the middle one ovate-acuminate, slightly lozenge-shaped ; crest thin, concave, lunate, rounded.

PHALÆNOPSIS *ROSEA ;* foliis oblongis coriaceis acutis apice recurvis, scapo cernuo ramoso tortuoso subclavato, floribus subcarnosis, sepalis ovatis, petalis ovalibus paulò latioribus, labello ascendente tripartito laciniis lateralibus lineari-spathulatis lunatis intermediâ ovatâ, cristâ lunatâ rotundatâ depressâ emarginatâ.

Phalænopsis rosea: *Lindley in Gardener's Chronicle,* 1848, p. 671, *with a woodcut; aliàs* Phal. equestris: *Reichenbach, jun., in Linnæa,* 1849, p. 865 ; *aliàs* Stauroglottis equestris: *Schauer in Act. Acad. Nat. cur.,* xix., suppl. 432.

THIS charming plant has found little favour among growers of Orchids, from their not knowing how to manage it. For the most part it appears in collections as a small tuft of broad inelegant leaves, throwing up now and then a puny scape of pallid flowers, in which there is scarcely an element of beauty. But the accompanying figure, which is a faithful representation of the plant as it is grown in Mr. Rucker's collection, shows that when the plant becomes old and healthy, and is perfectly well grown, it assumes quite another appearance. From amidst the deep green convex leaves springs up a branched scape, eighteen inches or two feet high, of an intense and shining purple, at the ends of which appear for months together a long succession of rosy star-like flowers, having a most brilliant ruby lip, warmed with yellow at the base, and enriched by intense violet at the upper end.

The history of the species was originally published in the *Gardener's Chronicle* of Oct. 7th, 1848, in the following words :—"This is a very unexpected addition to the genus Phalænopsis, of which it has exactly the habit. The flowers are small, numerous, and arranged in a loose spike. The lip wants the tendrils so remarkable in *Ph. amabilis* and *grandiflora*, instead of which it is bright rose-colour, with almost the shape of a trowel. It was found in Manilla by Mr. T. Lobb, who sent it to Messrs. Veitch, and who describes it as having a spike of flowers from twelve to eighteen inches long; that which we saw was not more than four inches long, but it had blossomed at sea, and is probably inferior to what it will become hereafter." How inferior, a comparison of this description with the present figure will amply show.

In the year 1849, the younger professor Reichenbach, who has distinguished himself by his critical acquaintance with Orchids, republished it in the *Linnæa*, a German botanical periodical, under the name of *Ph. equestris*, he having ascertained that a certain *Stauroglottis equestris*, described by M. Schauer in his account of the dried Orchids collected by Professor Meyer, was not distinguishable from Phalænopsis. We see no necessity for altering the name by which we originally made the plant known, even although the specific name *equestris* was applied to the plant at an earlier date. We, however, subjoin M. Reichenbach's specific character for the convenience of those botanists who do not possess the *Linnæa* :—

Phalænopsis equestris ; fo. oblongis, cuneatis, ped. ex axillis squamarum vetustarum exortis, teretiusculis, 3-vaginatis, nunc ramosis, superne flexuosis, br. minimis, acutis, p. ph. e. oblongis, acutis, p. ph. i. obovatis, acutis, lb. tripartito, partitionibus lateralibus lunatis, obtusis, divaricatis, intermedia oblonga apice in apiculum retusum attenuato, callo postice bilobo in basi.

We did not learn at Mr. Rucker's that any peculiar treatment was given to this plant, whose admirable health seems only owing to care, rest, and ample ventilation, combined with the ordinary requisites of skilful management.

446. DENDROBIUM ALBUM. *Wight.* A neat-looking Indian epiphyte, with pure white flowers. Introduced by Messrs. Veitch. (Fig. 226.)

The following is the account given by Dr. Wight of this species :—

" Erect, jointed ; stems enlarging from the base to the apex, internodes much shorter than the leaves. Leaves oblong, elliptic, acuminate. Flowers axillary, paired, long-peduncled ; sepals ovate, acute ; lateral ones falcate ; petals obovato-elliptic, obtuse, larger than the posterior sepal. Lip three-lobed ; lateral lobes entire, obtuse, middle one cucullate, ovate, acute, saccate at the base, ciliate. Flowers pure white. Native of the Iyamally Hills. Flowering in September. This is one of the handsomest of the genus I have yet met with ; the large pure white flowers and dark foliage are very conspicuous. It seems to be rather rare, as I have only once obtained specimens."—*Indian Orchids,* no. 1645.

The plant is scarcely distinct from *D. aqueum,* figured in the *Botanical Register,* 1843, t. 54 ; appearing to differ in nothing except a more narrow middle lobe of the lip, more distinct fringes upon its edge, and an absence of the green tinge which has been observed in *D. aqueum.*

447. CENTROSOLENIA PICTA. *Hooker.* A weedy-looking hot-house perennial with mottled leaves. Flowers whitish, with a hairy pink tube. Belongs to Gesnerads. Introduced at Kew.

Sent by Mr. Spruce from the banks of the Amazon. It is remarkable for its beautifully painted, blotched or mottled leaves. Its flowers are large and white, destitute of the long fringe to the limb so characteristic of *C. glabra,* and the opposite leaves are here nearly equal in size. A procumbent and creeping plant, growing in dense tufts. Stems branched, cylindrical, fleshy, downy. Leaves opposite, on long

226

terete footstalks, oval or obovate, rather fleshy, crenato-serrate, unequal in size, hirsutely velvety on both sides, penni-nerved and reticulated, the nerves very prominent beneath; above, many of the leaves are blotched with brown and paler green. Peduncles axillary, clustered, single-flowered, bracteated, shorter than the calyx. Bracteas linear, acu_minate. Calyx deeply five-partite, the segments lanceolate, acuminate, inciso-serrate, much shorter than the corolla. Corolla hirsute, large, white: the tube elongated, infundibuliform, running down at the base into a blunt spur: limb of five spreading, rounded lobes, obscurely crenated at the margin. A native of tropical America, and, like its allies, of a succulent, decumbent habit. It grows freely in a warm and moist atmosphere, such as is suitable for tropical Orchids. A mixture of light peat-soil and leaf-mould suits it. The pot or pan must be well drained; and during winter, an excess of moisture must be guarded against. It increases readily from cuttings, which root quickly if placed in a warm frame, without the aid of a bell-glass.—*Bot. Mag.*, t. 4611.

448. BEGONIA MARTIANA. *Link & Otto.* A tuberous greenhouse plant, with rich rose-coloured flowers. Native of Mexico. Flowers in the summer and autumn. (Fig. 227.)

Occasionally only we find this pretty plant among collections of Begonias. It was first procured for the Royal Garden of Berlin, and published by Link & Otto, from whom we borrow the following memorandum and the annexed cut:—

"The stem is branching, round, green, from three to four feet high. The long side of the leaf is nearly three inches long, and from one to two inches broad; the short side is scarcely one inch in length or breadth; the upper surface is dark green, the lower paler and shining. The teeth sometimes have a short point in front; the petiole is round; the panicle short and bears but few flowers. From the axils of the leaves grow small bulbs. The male flowers have four red petals, of which the larger are six lines in length and breadth, and the shorter from three to four lines long, and scarcely two broad. The female flowers have five petals of very unequal size. The seed vessel is furnished with three wings, of which two are narrow and one broadish; the upper one being obtuse. This species is closely allied to *Begonia incarnata*, but the leaves are differently cut at their edges, and quite smooth without ciliæ. The panicle also has fewer flowers. The tubers of this plant were sent to us from Mexico by M. Deppe. Its beautiful flowers last from July till September. Like all Begonias it likes a light soil of vegetable mould and loam mixed with river-sand; it may be kept in summer in a protected place in the open air, or in an open greenhouse. In autumn the plant dies down, and the tuber alone remains behind, which should be kept during winter in a temperature of from 45° to 50° Fahr. in a cold house until the spring, when it should be planted out in a hotbed, where it will soon strike root and flower. The plant may be propagated in various ways: 1st, by seeds sown in pots; 2nd, by cuttings, which easily take root; 3rd, by dividing the root; and lastly, by means of the little tubers in the axils of the leaves. These fall when the branches die, and may be kept during the winter in dry earth, and be in spring placed in a hotbed, where they soon take root and come up."
—*Link & Otto, Icones*, no. 25.

227

449. STENANTHIUM FRIGIDUM. *Kunth.* (*aliùs* Veratrum frigidum *Chamisso.*) A brown-flowered, half-hardy, herbaceous plant, belonging to the Order of Melanths. Native of Mexico, where it is called Cebadilla de tierra fria.

This is a perennial grassy-leaved plant, very much like a Tuberose in appearance before it comes into blossom. It is remarkable for the dull blackish-purple colour of its flowers, which appear in drooping panicled leafy racemes at the top of a leafy stem about 3 feet high. The only interest that attaches to this plant consists in its poisonous qualities. From the name under which it was sent home by Mr. Hartweg it may be supposed to furnish a part of the venomous Sabadilla seeds of commerce, from which Veratria is prepared. Dr. Schiede says (*Linnœa*, vol. iv., p. 226,) that the inhabitants of Mount Orizaba, where it grows wild, know it to be dangerous to the horses that bite it, and in another place that it is called the "Sevoeja." It is probably a hardy perennial requiring to be grown in peat soil and a rather moist situation during summer. It is increased freely by dividing the old roots when in a state of rest. It flowers in June and July, and obtains a height of 2 or 3 feet.—*Journal of Hort. Soc.*, vol. i.

450. ACACIA COCHLEARIS. *Wendland.* (*aliàs* Mimosa cochlearis *Labillardière.*) A handsome, hardy, greenhouse shrub, with balls of yellow fragrant blossoms, appearing in January and February. Native of the West Coast of New Holland. (Fig. 228.)

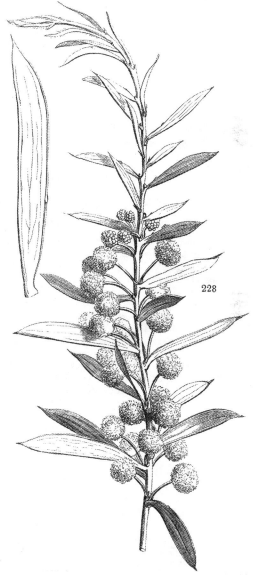

228

This very pretty species is one of those with the phyllodes ending in a sharp point, and the flowers in spherical heads. It is nearly allied to *A. lanigera*, from which it is distinguished by its shorter and less acuminate phyllodes, which are indistinctly marked by three (or perhaps four) parallel veins, instead of being so filled by a crowd of veins as to have a regularly striated appearance. In the cold Conservatory of the Horticultural Society it is a favourite object. The late lamented Mrs. Molloy, whose death was so grievous a loss to science, in a memorandum now before us says, " This is one of our handsomest shrubs, and renders the passing gale quite fragrant. About the Vasse River it exists in great numbers, forming thickets intermixed with Jacksonias. It has a rich profusion of flowers, and has long wreaths and garlands, studded with their blossoms, hanging from the tree, which usually attains the stature of 12 or 14 feet. It blossoms from September till the beginning of November, growing in black sand, and is most vigorous when there is partial moisture, as upon the banks of rivers."

451. SPHÆROSTEMA PROPINQUUM. *Blume.* (*aliàs* Kadsura propinqua *Wallich.*) A hothouse climbing shrub, with pale yellow fragrant flowers. Native of Nepal. Introduced at Kew. Flowers in June.

Dr. Wallich, to whom we are indebted for our plants, discovered the species on Sheopore and other hills at Lankoo, Nepal. Dr. Hooker found it frequent at from 7—9000 feet in Sikkim-Himalaya. It is a handsome and fragrant shrub : the natives eat the fruit, which consists of many berries attached to a receptacle : the latter elongates itself as the fruit advances to maturity, when the whole resembles a long bunch of red currants. A much branching, twiggy, somewhat climbing shrub, glabrous. Leaves alternate, on short petioles, ovate, much and finely acuminated, denticulate at the margin, penninerved, rather glaucous beneath. Peduncles longer than the petioles, axillary, single-flowered, solitary, or two to five or six in a fascicle, bearing several remote appressed acuminate bracteoles. Male flowers with nine sepals, arranged in threes ; the three outer smaller and calyciform, the six inner petaloid, coriaceous, rotundate, spreading, pale yellow, corolloid. Female flowers, according to Dr. Wallich, with sepals as in the male. Ovaries very small, numerous, fleshy, ovate, imbricated into a subglobose mass. Style none. Berries globose, fleshy, numerous, smooth, scarlet, two-seeded, arranged in a cylindrical spike, six inches long, with the rachis slightly compressed, muricated as it were by the numerous tubercles to which the berries were attached. Although not conspicuous as a showy flowering plant, yet the smooth leaves, general neat habit, and free growth of this species, make it worthy of cultivation. It is well adapted for training up rafters or on trellis-work. The plant from which the drawing was made is growing luxuriantly in light loam, and trained against the glass in the Palm-stove; it will also succeed in a warm green-house. It increases freely by cuttings placed under a bell-glass, and treated in the usual way.—*Bot. Mag.*, t. 4614.

452. CENTRADENIA OVATA. *Klotzsch.* A stove herbaceous plant, found by Warczewitz in Central America. Flowers pink. Belongs to Melastomads. Introduced by Mr. Mathieu, nurseryman, Berlin.

Stem and branches between spreading and erect, four-cornered, placed in four rows, and covered with close bristle-like hairs. Leaves opposite, triple-nerved, stalked, ovate, acute, bright green, smooth and shining above, with stiff bristles at the edges, pallid and downy on the veins beneath. Cymes terminal, trichotomous, many-flowered. Petals obovate, deep flesh-coloured. Sepals lanceolate, acute, closely ciliated.—*Gartenzeitung*, 1851, p. 354.

453. CENTRADENIA DIVARICATA. *Klotzsch.* A stove herbaceous plant, found by Warczewitz in Central America. Flowers white. Belongs to Melastomads. Introduced by Mr. Mathieu, nurseryman, Berlin.

Branches long, straggling, hoary. Leaves membranous, unequal sided, stalked, long, acuminate, shortly narrowed at the base, deep green above, closely bristly near the edge, pallid on the under-side, with downy veins; every other one small and persistent. Flowers few, terminal. Petals white, obovate. Sepals short, ciliated.—*Gartenzeitung*, 1851, p. 354.

454. NICOTIANA ALATA. *Link & Otto.* (*aliàs* ? N. decurrens *Agardh ; aliàs* ? N. persica *Lindley.*) A handsome tender annual, with fragrant white and green flowers. Native of South Brazil. Belongs to Nightshades. (Fig. 229.)

We translate literally the account given of this plant by Link and Otto. " The stem is from four to five feet in height, branching, with distant, glandular hairs. The leaves are from three to four or more inches long, and from one to two inches broad ; the upper ones are smaller ; they are all decurrent and form narrow wings on the stem, obtuse and with a small callous point, but a little repand at the edges and toothed, the teeth having also little callous points, on both sides rough with small somewhat closely pressed hairs, and at the edges furnished with distant and glandular hairs. The flowers are placed rather far apart from each other on a raceme ; the lower pedicels are one inch long, the upper ones are shorter. The rough calyx is not quite an inch in length, tubular ; its teeth are long and very narrow. The flowers are white and sweet-scented ; the tube from two to three inches long, a little expanding at the top ; the teeth of the limb, eight lines in length, are oval, somewhat expanded, obtuse. Stamens as long as the tube. Style somewhat longer. Capsules oblong. The seeds of this plant were sent by M. Sello in 1827 from Brazil. They should be sown in the spring in pots, and the seedlings should be planted out in the open ground when the frosts are gone. The plant is hardy and may be kept in winter in a temperature of from 38° to 43° Fahr., and as such plants as are strongest flower best and produce most seeds, they should be so treated. The soil should be light, but rich and mixed with sand. The large white odoriferous flowers, forming nice-looking tufts, render the plant suitable for bedding out. The flowers close in the day-time and hang down, but open at night. If the weather is cloudy they open as early as five P.M., but if clear not before six and a half P.M. ; in like manner they shut in the morning at six if the weather be clear, but not before seven if it be overcast."

Such is the account given by Link & Otto of a plant which we think is beyond all doubt what Sir Henry Willock found cultivated in Persia and sent to England as the source of Shiraz Tobacco ; in consequence of which it was called *N. persica* by one of us, and, according to M. Walpers, *N. decurrens*, by Bishop Agardh. We must, however, observe that the Persian plant was not observed to be a perennial ; nor do the leaves appear to have been so distinctly decurrent as is represented in the accompanying figure : but the specimens which have been preserved show that the leaves were somewhat decurrent, even near the summit of the flowering stem. This identification of plants supposed to be distinct leads to the inquiry of how a South Brazilian plant came to be cultivated in Persia as Tobacco ? and also whether any Brazilian Tobacco is manufactured from it ? We trust that some one will be able to answer these questions, as well as many others connected with the history of commercial Tobacco ; as, for instance, is any Havannah Tobacco prepared from *N. amplexicaulis*, as George Don reported ? Is the white-flowered Guatemala Tobacco a species distinct from the Red Virginian, *N. Tabacum* ? Are the red-flowered Tobaccos all varieties of *N. Tabacum* ? or do they belong to different species, as some pretend ? What yields the pitchy Tobacco of Latakia : or the mild Tobacco of Syria ? The Djebelé seems to belong to *N. Tabacum*. Is it true that *N. paniculata* is cultivated in the East ? How came *N. rustica* to be grown in Egypt and Tunis, where it produces the fragrant but strong Tombaki Tobacco, which was shown at the Great Exhibition of all Nations ? Of what country is *N. rustica* certainly a native ? All these are interesting questions, to not one of which we believe can a *satisfactory* answer be found in books. *N. alata* is lost in English gardens, but might perhaps be recovered from Berlin.

455. GRAELLSIA SAXIFRA-GÆFOLIA. *Boissier.* (*aliàs* Cochlearia saxifragæfolia *De Cand.*) A hardy herbaceous plant, with white flowers. Belongs to Crucifers. Native of the mountains of Persia.

This is a little plant, with long, kidney-shaped or roundish leaves, very coarsely notched, and smelling strongly of garlic. The flower-stems are about nine inches high, and bear a compound corymb of small white flowers resembling those of the common scurvy-grass. It has not produced any fruit. It is a hardy perennial, growing freely in any good rich garden soil, and well suited for planting on rockwork. It flowers in July and August, and is easily increased by dividing the old plants in autumn or spring, or by seeds: the plants raised from seed will not flower before the second season. It must be considered a good hardy plant for rockwork, and rather showy, as it flowers abundantly —*Journal of Hort. Soc.*, vol. i.

456. ALLIUM CASPIUM. *Bieberstein.* (*aliàs* Amaryllis caspia *Willdenow ; aliàs* Crinum caspium *Pallas.*) A green-flowered ugly bulb, from the deserts of Western Asia; flowering in May.

Native of the deserts of Astrachan and Tezzier. Dr. Stocks finds it in Scinde, and obligingly sent bulbs to the Royal Gardens, which flowered in May 1851. It has so little of the ordinary appearance of an Onion, that Willdenow called it an *Amaryllis*, and Pallas a *Crinum*. It has, however, all the characters of *Allium* and the same savoury odour.—*Bot. Mag.*, t. 4598.

457. FUCHSIA TETRADACTYLA. *Lindley.* A small scarlet-flowered species, from Guatemala.

A slender downy plant about two feet high, with very soft branches of a dull crimson colour. The leaves are opposite, about twenty-seven inches long, half of which belongs to their stalks, obovate-oblong, obtuse,

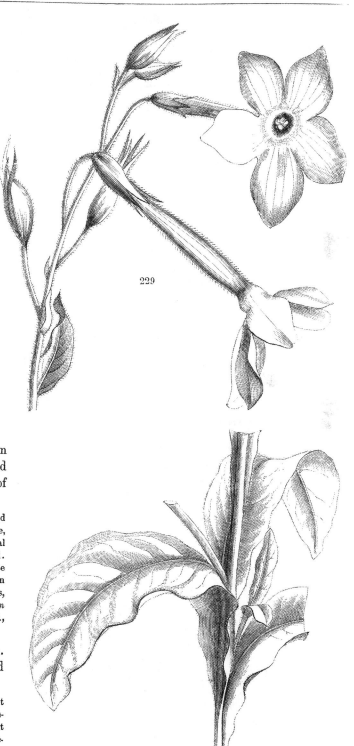

229

a little blistered in consequence of their lateral veins being much sunken. The flowers, which are small, and deep rose-colour, grow singly in the axils of the leaves. The calyx is half an inch long; the petals much shorter and paler than the lobes of the calyx, flat, blunt; and the stamens much shorter than the petals; the style longer than all, with a large star-shaped stigma divided into four fleshy finger-like rays. It is nearly allied to *F. thymifolia* and *cylindrica*. It appears to be a greenhouse plant. As yet it has been grown in sandy peat, but there is reason to believe that it will succeed under the same treatment as those numerous hybrids which are to be seen during the summer in every greenhouse. Like *F. fulgens* it has a large fleshy root, so that in autumn it may be stowed away in any corner, where it may be kept comparatively dry and free from frost till spring. Its flowers are, however, too small to render it interesting to any except botanists.—*Journal of Hort. Soc.*, vol. i.

458. GALEANDRA DEVONIANA. *Lindley.* A handsome terrestrial Orchid, from tropical America. Flowers cream-coloured and brown.

This was first detected by Schomburgk on the Rio Negro, a river which discharges itself into the Amazon; and Mr. Spruce has been so fortunate as to meet with it in the same locality, and we received a Wardian case from him in July of the present year 1851, containing the flowering specimen in excellent condition, which we here represent. Schomburgk saw it growing five to six feet high, and in clusters or patches from ten to twelve feet in circumference. Stems uniform to the base (no pseudo-bulbs), clustered, three to five or six feet high, scaly below, leafy above: leaves much sheathing at the base, linear-ensiform, acuminated, striated, glabrous, membranaceous. Panicle terminal, with few but large flowers; branches and peduncles bracteated. Sepals and petals spreading and slightly ascending, lanceolate, striated, darkish-purple, green at the margin and at the base externally. Lip very large, projecting, white, tipped and streaked with purple, broadly obovate, obscurely three-lobed, the sides meeting so as to form a lax tube around the column, intermediate or spreading, deflexed, retuse: near the base within are four lamellæ. Column within the tube-like portion of the labellum, slightly winged at the margin. Anther with a large, downy, erect crest. This is a tropical terrestrial Orchid, and therefore requires to be kept in a warm stove or Orchideous house. It may be potted in turfy peat-soil made rather firm in the pot, and well drained. In winter it must be so placed as not to suffer from excess of moisture, either in the atmosphere or in the soil.—*Bot. Mag.*, t. 4610.

459. HERMANNIA INFLATA. *Link & Otto.* A greenhouse shrub, with large inflated calyxes and short purple corollas. Said to be a native of Mexico. Belongs to Byttneriads. Introduced by the Berlin Garden. (Fig. 230.)

230

The stem of this is upright, from one to two feet high, and like the leaves, calyx, and seed-vessels, covered with thick stellate hairs. The leaves are one inch long, and from eight to ten lines broad, shortly pointed, finely notched, with five principal nerves, themselves branching into smaller nerves, somewhat wrinkled, very hairy, but still greenish. The petiole is from two to three lines long. The flowers proceed from the axils of the upper leaves, are bent downwards, and have a peduncle four lines long. The calyx is inflated, membranous, thinly covered with hairs, and has five short ovate teeth. The petals, five in number, are little longer than the calyx, and are of a dark purple red colour. The five stamens are composed of expanded filaments, growing a little together at their base, and of long anthers, extending to a point. The seed-vessels are five-cornered, have five furrows, and five cells, each containing many seeds. Styles five, united. We received the seeds of this plant from Temascaltepec, in Mexico, where they were collected by M. Deppe. It is the only Hermannia hitherto found out of South Africa, and there is no doubt that the plant is a Hermannia. It flowers with us in August and September, but has not yet produced any seeds. It likes, in summer, an open situation, not too much exposed to the sun, and in winter a temperature of from 50° to 54° Fahr. In winter the plant is tender, and requires a good place near the light. The soil should be half loam and half loam and river-sand. It may be easily propagated by cuttings.—*Link & Otto.*

We reproduce the account of this singular plant, in the hope that some one may find it among his collections. We have seen nothing like it from Mexico; but we can hardly distinguish it from a species of Hermannia found by Forbes at Algoa Bay.

460. COREOPSIS FILIFOLIA. *Hooker.* A handsome hardy annual, with bright yellow flowers. Native of Texas. Belongs to Composites. (Fig. 231.)

Although introduced to this country by the late Mr. Drummond, almost twenty years ago, this plant is still scarcely known in cultivation. It has, however, been lately rescued from oblivion by the French, and richly deserves a place in all gardens of hardy flowers. The stems grow about three feet high, and bear a profusion of rich golden-yellow flowers with a crimson disk. " Of all the narrow and divided-leaved species of Coreopsis," says Sir William Hooker, " this has unquestionably the narrowest foliage, and which, if examined carefully, exhibits the most fleshy texture, the underside semi-terete and presenting no appearance of a nerve or costa, which indeed is only indicated on the upper side by the presence of a furrow. Its nearest ally is perhaps *C. tenuifolia* ; but there, besides the difference in foliage, the disk is described as being of the same colour as the ray, and the florets of the ray much narrower."

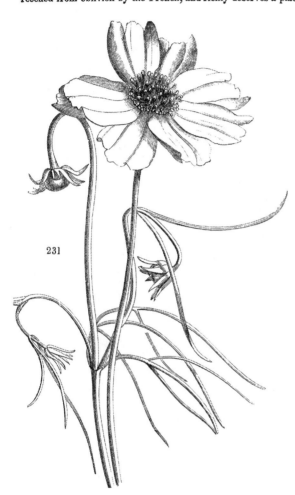

231

461. POTENTILLA AMBIGUA. *Cambessèdes.* A handsome prostrate perennial, with fine yellow flowers. Native of the Himalaya. Belongs to Roseworts. Introduced at Kew.

A well-marked, hardy, Himalayan species of Potentilla, with a compact habit and large yellow flowers, produced abundantly during the summer months. Jacquemont detected it in fissures of rocks in Kanaor, near Rogui, elev. 9000 feet, in about lat. 32°, long. E. 78½°, where it was likewise found by Capt. Henry Strachey ; thence it appears to extend eastward through Nepal to Sikkim-Himalaya, where it was found by Dr. Hooker in woods at an elevation of from 12-13,000 feet above the level of the sea. Its nearest affinity is with *P. eriocarpa* Wall. ; but there the stem is scarcely leafy, and the leaflets are longer and much more divided. From a woody perennial root, many closely-placed stems diverge : they are ascending, six inches to a foot long, frequently purple, leafy, clothed with soft silky hairs, as is, more or less, every part of the plant. Leaves on longish petioles (which have two large, ovate usually entire stipules at the base), ternate ; leaflets cuneato-obovate, trifid at the apex, of a firmish texture, glaucous beneath, the lateral ones sessile, the terminal one on a short petiolule. Peduncles slender, terminal, single-flowered. Flowers large, yellow. Calyx with five large, obovate, spreading bracteas, glaucous beneath, entire. Petals large, rather obcordate than obovate. A native of the elevated regions of the Himalaya, and sufficiently hardy to endure the cold of this climate during the last winter. Till it has stood the test of a severe winter, however, it may be desirable to keep a few plants in pots under protection, for, being of a suffruticose habit, it may probably suffer from severe frost. It is a free-growing species, increasing rapidly by its stoloniferous roots, and soon forming a large patch. It continues to flower until late in the autumn.—*Bot. Mag.*, t. 4613.

462. VACCINIUM ROLLISONI. *Hooker.* An evergreen greenhouse bush, with red flowers and short blunt leaves. Native of Java. Introduced by Messrs. Rollison.

From the collection of Messrs. Rollison, Tooting Nursery, where it produced its rich scarlet flowers in August, 1851. Introduced by their collector, who found it growing on the lava of the " silent volcanoes " of Java, on the highest land in the island. We have specimens of the same from Salak mountain, Java, from Mr. Thomas Lobb. It forms a handsome evergreen bush, with glossy Box-like leaves, and what is wanted in the number of flowers, is compensated by their beauty of colour. It does not appear to be anywhere described, either under Vaccinium or Agapetes. It is not *Agapetes microphylla* Junghuns, for that has leaves three to four inches long. Requires to be treated as a greenhouse plant. In the summer it may be placed in the open air in a shady place. Like the rest of this tribe of plants, it thrives in light sandy peat-soil, and is readily increased by cuttings.—*Bot. Mag.*, t. 4612.

463. NYMPHÆA SCUTIFOLIA. *De Candolle.* (*aliàs* N. capensis *Thunberg.*) A most beautiful hothouse aquatic, with large blue flowers. Native of the Cape of Good Hope. (Fig. 232.)

We have two blue Nymphæas in cultivation, both called *N. cœrulea.* Of them, one, which is very common, is a native of the Nile, and has leaves without indentations, and small flowers: the other, found at the Cape of Good Hope, has flowers four times as large, with four times as many petals and stamens, and leaves with coarse indentations at the edge ; this is as much more rare as it is more beautiful ; it is sometimes called *N. cyanea* in Gardens. Concerning the wild habits of the last, now figured after a beautiful design in Van Houtte's *Flore des Serres et des Jardins,* we have little information ; Dr. Harvey says it is the only Cape water-lily, and is found in various parts of the colony ; Thunberg gives the streams at Lange Kloof. What we certainly know is, that it is by very far the handsomest blue water-lily that we possess.

232

INDEX OF VOLUME II.

[*Plate* signifies the coloured representations; *No.* the number of the Gleanings and Memoranda; *fig.* the woodcuts.]

END OF VOLUME II.

LONDON:
BRADBURY AND EVANS, PRINTERS, WHITEFRIARS.

Printed in the United States
By Bookmasters